COMPUTER SCIENCE, TECHNOLOGY AND APPLICATIONS

Additional books in this series can be found on Nova's website under the Series tab.

Additional E-books in this series can be found on Nova's website under the E-book tab.

COMPUTER SCIENCE, TECHNOLOGY AND APPLICATIONS

COMPUTER NETWORKS, POLICIES AND APPLICATIONS

CAROLYNN M. BERGER
AND
ANGELA D. STREUSSE
EDITORS

Nova Science Publishers, Inc.
New York

For permission to use material from this book please contact us:
Telephone 631-231-7269; Fax 631-231-8175
Web Site: http://www.novapublishers.com

Additional color graphics may be available in the e-book version of this book.

LIBRARY OF CONGRESS CATALOGING-IN-PUBLICATION DATA

Computer networks, policies, and applications / editors, Carolynn M. Berger and Angela D. Streusse.
p. cm.
Includes index.
ISBN 978-1-61209-088-7 (hardcover)
1. Computer networks. 2. Telecommunication--Government policy--United States. 3. Internet--Social aspects--United States. I. Berger, Carolynn M. II. Streusse, Angela D.
TK5105.5.C645 2011
004.6--dc22 2010047619

Published by Nova Science Publishers, Inc. † New York

CONTENTS

PREFACE

This book presents current research in the study of computer networks, congressional policies and issues, as well as various applications and technologies powered by computers. Topics discussed include internet and computer-mediated technologies as new tools in the study of shyness; broadband internet access and the digital divide; bundling residential telephone, internet and video services; gender differences in high school internet coverage and ad hoc networks.

Chapter 1 - Debate has begun about what statutory and regulatory framework is most likely to foster innovation and investment both in physical broadband networks *and* in the applications that ride over those networks. Perhaps the most contentious element in that debate is whether competitive marketplace forces are sufficient to constrain the broadband network providers from restricting independent applications providers' access to their networks in a fashion that would harm consumers and innovation.

The telephone and cable companies are deploying wireline broadband networks with unique architectures. For example, Verizon is deploying optical fiber all the way to the customer premise, while AT and T is deploying fiber to a node and then using DSL over existing copper lines to reach the customer premise, and Comcast and other cable companies are deploying a hybrid fiber-coaxial cable network. But in each case, their broadband networks have the same basic structure, with three primary components — the broadband "last mile" grid to end-user customers, the company's proprietary IP network, and the company's facilities in what has traditionally been called the internet backbone (and is often referred to as the "public internet"). This chapter analyzes these three components to identify the parameters that network providers have within their control (such as their choices about network architecture, overall bandwidth capacity, bandwidth reserved for their own use, traffic prioritization, the terms and rates for access to their networks and for their retail services) that can affect how end users and independent applications providers can access their networks, how those parameters contribute to the management and operation of the network, and how those parameters might be used strategically to harm competition for, and consumers of, voice over internet protocol (VoIP), video, and other applications that ride over broadband networks.

The report then reviews various legislative proposals affecting network access to assess their potential impact on broadband network providers' ability to manage their networks and to practice anticompetitive strategic behavior. Two bills, H.R. 5252 as passed by the House and H.R. 5252 as amended by the Senate Commerce Committee (originally introduced as S.

2686), specify particular consumer rights to broadband access. Three bills — H.R. 5273, S. 2360, and S. 2917 — propose variations on "network neutrality" rules that have provisions affecting the access of independent applications providers, as well as consumers, to broadband networks. Two other bills, H.R. 5417 and S. 2113, propose modification of existing competition law (involving antitrust and unfair methods of competition) to explicitly address broadband access issues.

Chapter 2 - The "digital divide" is a term that has been used to characterize a gap between "information haves and have-nots," or in other words, between those Americans who use or have access to telecommunications technologies (e.g., telephones, computers, the Internet) and those who do not. One important subset of the digital divide debate concerns high-speed Internet access, also known as *broadband*. Broadband is provided by a series of technologies (e.g. cable, telephone wire, fiber, satellite, wireless) that give users the ability to send and receive data at volumes and speeds far greater than current "dial-up" Internet access over traditional telephone lines.

Broadband technologies are currently being deployed primarily by the private sector throughout the United States. While the numbers of new broadband subscribers continue to grow, studies conducted by the Federal Communications Commission (FCC), the Department of Commerce (DOC), and the Department of Agriculture (USDA) suggest that the rate of broadband deployment in urban and high income areas may be outpacing deployment in rural and low-income areas.

Some policymakers, believing that disparities in broadband access across American society could have adverse economic and social consequences on those left behind, assert that the federal government should play a more active role to avoid a "digital divide" in broadband access. One approach is for the federal government to provide financial assistance to support broadband deployment in underserved areas. Others, however, believe that federal assistance for broadband deployment is not appropriate. Some opponents question the reality of the "digital divide," and argue that federal intervention in the broadband marketplace would be premature and, in some cases, counterproductive.

Legislation introduced (but not enacted) in the 109th Congress sought to provide federal financial assistance for broadband deployment in the form of grants, loans, subsidies, and tax credits. Many of these legislative proposals are likely to be reintroduced into the 110th Congress. Of particular note is the possible reauthorization of the Rural Utilities Service (RUS) broadband program, which is expected to be considered as part of the 2007 farm bill. Legislation to reform universal service – which could have a significant impact on the amount of financial assistance available for broadband deployment in rural and underserved areas – has been introduced into the 110th Congress (H.R. 42, S. 101).

In assessing such legislation, several policy issues arise. For example, is the current status of broadband deployment data an adequate basis on which to base policy decisions? Is federal assistance premature, or do the risks of delaying assistance to underserved areas outweigh the benefits of avoiding federal intervention in the marketplace? And finally, if one assumes that governmental action is necessary to spur broadband deployment in underserved areas, which specific approaches, either separately or in combination, are likely to be most effective?

Chapter 3 - A prerequisite for the mass-market deployment of multi-hop ad hoc technologies is the capability of cooperating with existing wired infrastructure networks and providing transparent access to the Internet. However, current solutions to support connectivity between ad hoc networks and the Internet are based on complex mechanisms, such as Mobile-IP and IP tunnelling. In this chapter, the authors describe a lightweight and efficient architecture to build multi-hop hybrid ad hoc networks (i.e., using both wired and wireless links) employing proactive routing protocols, which can be used as flexible and low-cost extensions of traditional wired LANs. The authors' proposed architecture provides transparent global Internet connectivity and self-configuration to mobile nodes, without requiring configuration changes in the pre-existing wired LAN and relying only on basic layer-2 functionalities. The authors have prototyped the core components of this architecture for OLSR-based ad hoc networks. In this chapter, the authors present several experimental results such as to evaluate the network performance, and to determine an "optimal" setting of the routing protocol parameters.

Chapter 4 - Today's academic libraries are forced to enter into a significant number of licensing agreements every year for access to electronic resources. Vendors, or licensors, have the power to dictate the terms and conditions set forth in the license agreement. While libraries may attempt to negotiate and change certain terms, their attempts are often futile or met with strong resistance. Ultimately, the terms of these license agreements dictate the ability of libraries to provide access to electronic resources for their patrons. The license terms may limit or curtail access by certain patron groups, including alumni, walk-in users, the public, and other higher education institutions. The license agreement may also eliminate or restrict libraries' abilities to rely on the fair use provisions or engage in interlibrary loan under copyright law. Additionally, libraries may choose to or be forced for economic reasons to discontinue their print collection of titles that are also available electronically. These libraries may be left scrambling to provide access to these same resources if the licensor eliminates or changes its titles, years of coverage, or ceases to exist as an entity. As a result of these scenarios, libraries may find themselves unable to provide the level of intended access to electronic resources in accordance with their goals or mission statement. This inability to provide access can be directly tied back to the license agreement and its terms and conditions.

Chapter 5 - Technological advances and deregulatory actions now allow consumers to obtain their local and long distance telephone services, their high-speed Internet services, and their video services from competing technologies. The convergence of previously distinct markets has required companies to seek strategies for holding on to their traditional customers while seeking new ones. One of those strategies is for companies to offer bundles of "traditional" and "new" services at a single price that often represents a discount off the sum of the prices of the individual services. These bundled service offerings are favored by many consumers. They provide the convenience of "one stop shopping" and in some situations, by providing the full panoply of services at a fixed price, make it easier for consumers to comparison shop. They also are favored by many providers because they tend to reduce "churn" - the rate at which customers shift to competitors - and allow providers to exploit economies of scope in marketing.

But bundling also can create public policy issues for Congress. The bundled offerings typically provide some combination of interstate telecommunications services, intrastate telecommunications services, and non-telecommunications services (information services, video services, and even customer premises equipment) for a single price. The federal Universal Service Fund - the federal subsidy program that assures affordable telephone rates for high-cost (rural) and low-income telephone customers as well as for schools, libraries, and rural health facilities - is supported by an assessment on interstate telecommunications revenues only. But it is difficult to identify the portion of revenues generated by a bundled service offering attributable to the interstate telecommunications portion of that bundle. There is no unambiguous way for providers to assign a portion of the bundled price to interstate telecommunications services or for fund administrators to audit that assignment. In addition, some taxes are assessed upon one or more, but not all, of the services included in various bundled service offerings. This creates the same assessment and auditing problem for these taxes as exists for the federal Universal Service Fund. This has important policy implications at a time when many Members of Congress seek to shelter Internet services-which often are included in these bundles- from taxation without placing any group of providers at a competitive advantage or disadvantage.

Some observers have been concerned that bundled service offerings could have anticompetitive consequences if they foster industry consolidation or if a provider has market power for one of the services in its bundled offering and can use that offering to tie that service to a competitive service in a fashion that reduces competition for the competitive service.

Leaders in both the House and the Senate Commerce Committees have announced that in the 109" Congress they plan to review and reform the 1996 Telecommunications Act (P.L.104-104) in light of the market convergence that underlies the trend toward bundling.

Chapter 6 - The 9/11 Commission, in its recent report on the attacks of September 11, 2001, criticized the U.S. Intelligence Community's (IC) fragmented management structure and questioned whether the U.S. government, and the IC, in particular, is organized adequately to direct resources and build the intelligence capabilities that the U.S. will need to counter terrorism, and to address the broader range of national security challenges in the decades ahead.

The Commission made a number of recommendations, one of which was to replace the current position of Director of Central Intelligence with a National Intelligence Director (NID) who would oversee national intelligence centers on specific subjects of interest – including a National Counterterrorism Center (NCTC) – across the U.S. government, manage the national intelligence program, oversee the agencies that contribute to it, and have hiring, firing and budgetary authority over the IC's 15 agencies. The Commission recommended that the director, and the NCTC, be located in the Executive Office of the President and that a deputy NID be established to oversee the day-to-day operations of the Central Intelligence Agency (CIA). The Commission's recommendation to strengthen management authority over the IC is the latest contribution to an IC structural reform debate that dates at least to 1955, when arguments for stronger IC authority began to surface. OMB deputy director James Schlesinger in 1971 first broached the DNI concept.

Recently some Members of Congress have introduced intelligence community reform legislation that would establish the position of Director of National Intelligence (DNI), or strengthen DCI authorities.

Reactions to the concept of a DNI have been mixed since its inception. Supporters argue that the DCI cannot manage the IC, the CIA and serve as the President's chief intelligence advisor, and do justice to any of the jobs. Other than the CIA, the DCI also lacks hiring, firing and budget authority. They argue that the absence of strong, centralized leadership has resulted in divided management of intelligence capabilities; lack of common standards and practices across the foreign-domestic intelligence divide; structural barriers that undermine the performance of joint intelligence work; and a weak capacity to set priorities and move resources.

Opponents counter that a DNI would lose day-to-day control over the CIA, a natural power base and, as a result, influence. They also contend that placing the intelligence director in the Executive Office of the President, as the 9/11 Commission has proposed, risks the politicization of intelligence, giving the White House more direct control over covert operations, blurring the line between foreign and domestic operations and possibly shifting too much influence over intelligence to the Department of Defense. With regard to DOD influence, other opponents argue that a national director will shift the balance of control away from DOD, risking intelligence support to the warfighter. The congressional role includes deciding whether to establish the position of the DNI and its authority.

Chapter 7 - To improve the organization and capabilities of the U.S. intelligence community, the National Commission on Terrorist Attacks Upon the United States (9/11 Commission) recommended, among other things, that two entities be established — a National Intelligence Director (NID) and a National Counterterrorism Center (NCTC). The commission said that the NID should be located within the Executive Office of the President (EOP), and said the NCTC should report to the NID. Legislation designed to create these entities would place them either within the EOP (e.g., H.R. 5024 and H.R. 5050) or within an independent entity outside of the EOP (e.g., S. 2774 and H.R. 5040).

Questions have been raised regarding a number of issues associated with the establishment of the NID and the NCTC. One set of such questions centers on whether these proposed entities would be covered by various "general management laws" — broad statutes designed to regulate the activities, procedures, and administration of all or most executive branch agencies in such areas as regulatory and information management, financial management, procurement, and strategic planning. Earlier in 2004, CRS published a compendium of these general management laws and a companion report on major themes from these laws, with possible management policy options (CRS Report RL30795, *General Management Laws: A Compendium*; and CRS Report RL32388, *General Management Laws: Major Themes and Management Policy Options*).

This chapter examines whether more than 50 general management laws (e.g., the Administrative Procedure Act, the Inspector General Act, and the Competition in Contracting Act) would cover the NID and the NCTC under two scenarios: (1) if the entities were established within the EOP, and (2) if the entities were established as or within independent entities in the executive branch. The analysis focuses on the *coverage* of the selected management laws, not on whether the laws are likely to *affect* or *apply* to the NID or the NCTC. Also, some of the laws allow exemptions or contain other provisions that can affect their coverage.

In general, the report indicates that more of these management laws would appear to cover the NID and NCTC if they were created as independent entities in the executive branch than if they were placed within the EOP. In many cases, it was unclear whether the statutes' definitions of covered "agencies" included entities within the EOP. The report also indicates that Congress, if it so desires, could indicate that the NID and the NCTC would be covered by these management laws by either (1) amending the text of each of the management laws or (2) delineating in the statute creating the entities the particular management laws that should cover them.

Chapter 8 - ICT are challenge for the Tunisian economy. This study highlights the acceptance as a social issue that can helps to understand the behaviour of student toward ICT. A sample of 200 Tunisian students was used for the survey. The results show the links between motivator factors and acceptance and on the other hand some ties between acceptance, motivator factors and the antecedents, especially gender, age and study level. According to the authors' results some factors seem to be more important for students such as the perceived usefulness and the perceived enjoyment. Social pressure and perceived ease of use are considered as weak factor to understand the acceptance, exception for student between 20 and 22 years; the results show some differences.

Chapter 9 - Federal agencies collect and use personal information for various purposes from information resellers—companies that amass and sell data from many sources. GAO was asked to testify on its April 2006 report on agency use of reseller data. For that report, GAO was asked to determine how the Departments of Justice, Homeland Security, and State and the Social Security Administration used personal data from resellers and to review the extent to which agencies' policies and practices for handling this information reflected the Fair Information Practices, a set of widely accepted principles for protecting the privacy and security of personal data. GAO was also asked to provide an update on the implementation status of its recommendations and to comment on provisions of the proposed Federal Agency Data Protection Act. In preparing this testimony, GAO relied primarily on its April 2006 report.

Chapter 10 - Internet and computer technologies allow for the investigation of traditional questions in personality and experimental social psychology regarding person-context interactions in new ways. In this paper, the authors review recent survey- and laboratory-based studies that have used the Internet and computer technologies to study shyness. The authors also discuss how these technologies can be used as laboratory tools to manipulate context in the study of shyness and social behavior. The authors conclude with suggestions for future work using computer technologies in both basic and applied areas of research on shyness and social anxiety.

Chapter 11 - More women and girls are participating in athletics than ever before, however, media coverage of female athletes still lags behind that of male athletes (Pederson and Whisenant, 2002). The purpose of this study was to continue to assess media coverage by gender at the high school level by assessing the coverage of school websites. An initial check at the midpoint of the season evaluated six items, including the presence of coaches' names and rosters. Data from eight dates were then collected for the rest of the season to assess game results posted online. Results indicated that males received more coverage than females in general information and in game results. The findings point to potential Title IX violations by high schools.

Versions of these chapters were also published in *Internet Journal*, Volume 2, Numbers 1/2 and 3, published by Nova Science Publishers, Inc. It was submitted for appropriate modifications in an effort to encourage wider dissemination of research.

In: Computer Networks, Policies and Applications
Editors: C. M. Berger and A. D. Streusse
ISBN: 978-1-61209-088-7

Chapter 1

ACCESS TO BROADBAND NETWORKS*

Charles B. Goldfarb

ABSTRACT

Debate has begun about what statutory and regulatory framework is most likely to foster innovation and investment both in physical broadband networks *and* in the applications that ride over those networks. Perhaps the most contentious element in that debate is whether competitive marketplace forces are sufficient to constrain the broadband network providers from restricting independent applications providers' access to their networks in a fashion that would harm consumers and innovation.

The telephone and cable companies are deploying wireline broadband networks with unique architectures. For example, Verizon is deploying optical fiber all the way to the customer premise, while AT and T is deploying fiber to a node and then using DSL over existing copper lines to reach the customer premise, and Comcast and other cable companies are deploying a hybrid fiber-coaxial cable network. But in each case, their broadband networks have the same basic structure, with three primary components — the broadband "last mile" grid to end-user customers, the company's proprietary IP network, and the company's facilities in what has traditionally been called the internet backbone (and is often referred to as the "public internet"). This chapter analyzes these three components to identify the parameters that network providers have within their control (such as their choices about network architecture, overall bandwidth capacity, bandwidth reserved for their own use, traffic prioritization, the terms and rates for access to their networks and for their retail services) that can affect how end users and independent applications providers can access their networks, how those parameters contribute to the management and operation of the network, and how those parameters might be used strategically to harm competition for, and consumers of, voice over internet protocol (VoIP), video, and other applications that ride over broadband networks.

The report then reviews various legislative proposals affecting network access to assess their potential impact on broadband network providers' ability to manage their networks and to practice anticompetitive strategic behavior. Two bills, H.R. 5252 as passed by the House and H.R. 5252 as amended by the Senate Commerce Committee (originally introduced as S. 2686), specify particular consumer rights to broadband

---

* This is an edited, reformatted and augmented edition of a United States Congressional Research Service publication, Report RL33496, dated August 31, 2006.

access. Three bills — H.R. 5273, S. 2360, and S. 2917 — propose variations on "network neutrality" rules that have provisions affecting the access of independent applications providers, as well as consumers, to broadband networks. Two other bills, H.R. 5417 and S. 2113, propose modification of existing competition law (involving antitrust and unfair methods of competition) to explicitly address broadband access issues.

INTRODUCTION

Debate has begun about what statutory and regulatory framework is most likely to foster innovation and investment both in physical broadband networks *and* in the applications that ride over those networks. Perhaps the most contentious element in that debate is whether competitive marketplace forces are sufficient to constrain the broadband network providers from restricting independent applications providers' access to their networks in a fashion that would harm consumers and innovation. Or is government intervention needed in the form of what has been referred to as "network neutrality," unfair competitive practices, or other nondiscrimination rules placed on the network providers?

This debate has been stimulated by some fundamental changes in the telecommunications market environment — several technology-driven, several market-driven, and one regulatory-driven.

- Digital technology has reduced the costs for those firms that already have single-use (for example, voice or video) networks to upgrade their networks in order to offer multiple services over their single platform. The cost for these previously single-service providers to enter new service markets has been significantly reduced,[1] inducing market convergence. Most notably, cable companies are upgrading their networks to offer voice and data services as well as video services, and telephone companies are upgrading their networks to offer video and data services as well as voice services.
- Despite these lower entry costs, however, wireline broadband networks require huge sunk up-front fixed capital expenditures. This may limit the number of efficient broadband networks that can be deployed in any market to two (the cable provider and the wireline telephone company) unless a lower cost alternative becomes available using wireless or some other new technology.[2]
- Although wireless technology may provide a third or even fourth alternative, it is not likely to be a ubiquitous option anytime soon.[3] The commercial mobile wireless (cellphone), WiFi, and WiMAX technologies still require significant further technical developments before they will be able to provide comparable service and operate at the necessary scale. Moreover, spectrum is just being made available for these technologies, and in many cases parties currently using that spectrum must be moved to other spectrum.
- The new broadband networks are able to deliver potentially highly valued services, such as voice over internet protocol (VoIP) and video over internet protocol (IP Video), that are qualitatively different than most of the services that have been provided over the internet in the past. Where services such as e-mail and website searches are not sensitive to "latency" — the amount of time it takes a packet of data to travel from source to destination — these new services are sensitive to delays in

the delivery of packets of bits due to congestion or other problems.[4] As a result, the traditional internet "best effort" standard that does not guarantee that delays will not occur may be insufficient to meet customers' service quality requirements for these new latency-sensitive services.[5] More intensive network management may be needed to meet these quality of service (packet delivery) requirements.

- Equipment is being deployed in the broadband networks that can identify both the source of individual packets and the application to which individual packets are being put. With this equipment, network providers can give some packets higher priority than others, which can ensure that specific quality of service requirements are being met, but also could be abused to discriminate for or against particular applications or applications providers.
- Some new applications place very substantial bandwidth demands on the public internet and proprietary IP networks. For example, one industry analyst estimated that one particular application, BitTorrent software that uses file-sharing technology to download movies and other content, accounted for as much as 30% of all internet traffic at the end of 2004, and that peer-to-peer (P2P) applications, in general, represented 60% of internet traffic.[6] BitTorrent has been used both for legitimate purposes and for the illegal downloading of copyrighted materials, but has now been accepted by some mainstream content providers. For example, Warner Brothers has announced plans to make hundreds of movies and television shows available for purchase over the internet using BitTorrent software.[7] Other major industry players, such as Microsoft and Sony, have introduced movie download services that use P2P technology.
- Although the telephone and cable companies are deploying different network architectures,[8] they are pursuing business plans and regulatory strategies with the same key elements:
- They expect latency-sensitive video and voice services to be the "killer applications" that will generate the revenues needed to justify upgrade and buildout of their physical broadband networks.
- To minimize customer churn[9] and to gain an advantage over providers of single services, they market bundles of voice, data, and video services, with discounts that are greater the greater the number of services purchased. (It is expected by many that this "triple-play" bundle will be expanded to a "quadruple-play" bundle with the addition of mobile wireless service.)
- The set of services the telephone and cable companies plan to offer over their networks, despite having interactive components, follow the model of the customer being primarily a recipient of information, not a transmitter of information. Therefore the broadband network architecture they all are deploying is asymmetric — with significantly greater bandwidth available from the broadband provider to the customer than in the reverse direction.
- The video and voice services they offer, as well as other end-to-end services they plan to offer in the future, require quality of service assurances that they claim are not available on the "public internet," but can be provided on their proprietary IP networks. In order to assure the quality of service of their own offerings, the broadband network providers all seek to manage bandwidth usage on their proprietary broadband networks by reserving a significant proportion of their

network capacity for their own applications and by controlling the access that independent applications providers have to those networks through a variety of means, including charges for priority access.

- The Federal Communications Commission ("FCC" or "Commission") ruled in 2002 that cable modem service offered by cable companies, despite having a telecommunications component, is an information service and therefore not subject to the common carrier regulations imposed on telecommunications services in Title II of the Communications Act.[10] The FCC decision was upheld by the Supreme Court in June 2005.[11] Subsequently, the FCC ruled that DSL service offered by cable companies also is an information service.[12] As a result, neither cable modem service nor DSL service is subject to the interconnection, nondiscrimination, and access requirements of Title II.

Independent applications providers have voiced concern that the broadband network providers could abuse that control over network access to constrain — or entirely exclude — them from competing in the provision of applications, thereby undermining their ability to bring innovative applications to consumers. Some applications providers therefore have proposed enactment of statutory and regulatory requirements, such as nondiscriminatory access to broadband networks or network neutrality requirements. Others have been less confident about the ability to craft effective nondiscrimination or neutrality rules. They have suggested that government policy that promotes entry by broadband network providers that do not share the business plans of the cable and telephone companies might be a more effective way to foster innovation and investment in applications.[13] This might include prohibiting restrictions on municipal deployment of broadband networks, expediting the availability of spectrum for wireless broadband networks, and limiting the amount of such spectrum that can be acquired by companies owned by or in other ways affiliated with the wireline broadband providers. Current broadband network providers respond that, given existing market forces, they have neither the incentive nor the ability to constrain independent applications providers, that constraining their ability to manage their networks would discourage their investment in broadband networks, and that municipal networks enjoy an unfair advantage in capital markets.

To date, the debate has proceeded on an abstract level. The purpose of this chapter is to provide a more concrete discussion of access to wireline broadband networks. To that end, this chapter provides a discussion of what broadband networks look like; how both consumers and independent applications providers gain access to these networks; and the parameters available to network providers (such as their choices about network architecture, overall bandwidth capacity, bandwidth reserved for their own use, traffic prioritization, the terms and rates for access to their networks and for their retail services) that can affect end users' and independent applications providers' access to those networks.

What Do Wireline Broadband Networks Look Like?

The cable and telephone companies' networks are not static; they continue to be upgraded. The various network providers are in different stages of deploying digital IP technology. They are each employing unique network architectures that build off their legacy networks and/or reflect their perceptions of the most cost-effective networks to deploy. Despite these differences, however, each of the networks has three primary components, as shown in Figure 1, a schematic representation of a wireline IP broadband network. These three components are the broadband "last mile" grid out to end-user customers; the company's proprietary IP network, with servers for the various applications feeding into a service router that controls the flow of traffic all the way to the customer premise; and the company's facilities in what has traditionally been called the internet backbone (and is sometimes referred to as the "public internet"). The latter connects to independent applications providers, though it is also possible for independent applications providers to connect directly to a broadband network provider's proprietary IP network, as shown in Figure 1. Many independent applications providers also have substantial internet facilities. For example, in order to minimize the number of times their content must be handed off from one internet backbone provider to another when responding to an end-user query, these applications providers "cache" their content close to their customers by maintaining multiple servers scattered around the country in which they maintain frequently-updated databases. Since it is possible that congestion could cause delay at any of those handoff points, caching data at multiple servers reduces the risk of service degradation.

The schematic representation in Figure 1 may help elucidate a number of policy-related discussions. For example, it may help discussants visualize where and how end users and independent applications providers gain access to the broadband network; where and how congestion occurs that threatens the quality of latency-sensitive services; and how a network provider's capacity, architecture, prioritization, and service offering decisions could affect independent applications providers.

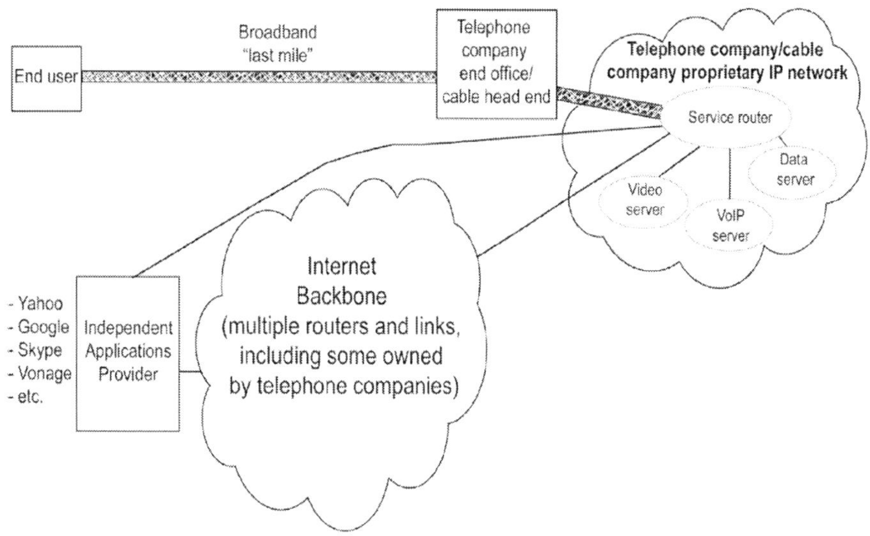

Source: CRS from multiple sources.

Figure 1. Schematic Representation of a Wireline IP Broadband Network.

The Broadband Last Mile

A wireline broadband network provider can choose among various network architectures for its last mile grid. A telephone company can choose to deploy optical fiber from its end office switch all the way to the home, or to the curb, or to a neighborhood node. If it brings the fiber to the curb or to a node, it can then complete the connection to the customer premise by attaching digital subscriber line (DSL) modems to the existing copper line running into the premise. For example, Verizon's Fios service deploys optical fiber all the way to the customer premise, while AT and T's Project Lightspeed deploys fiber to a neighborhood node and then uses existing copper lines and DSL modems to reach the customer.[14] Cable companies most often use hybrid fiber-coaxial cable (HFC) technology, deploying optical fiber from the cable company's head-end facility to a node and using coaxial cable from the node to the end-user premise. The fiber to the home architecture is much more costly to deploy, but can provide substantially more bandwidth than can be provided over a fiber/DSL or HFC last mile[15] and can have its bandwidth expanded more cheaply and easily as demand grows.

If a broadband network provider intends to offer multiple channel video service, it can choose between an architecture that "broadcasts" the signals of all the channels to the end-user premise (the cable company and Verizon approach) and an architecture that transmits to the end user only the particular video channel selected by the customer using her IP set-top box (the "call-up" approach used by AT and T). The "broadcast" approach requires more bandwidth.

Network providers have discretion over several other network parameters. For example, both the telephone and the cable companies have chosen to deploy asymmetric broadband networks that have far more bandwidth for the download of information to end-user customer premises than for the upload of information from end users. This architecture favors the development of applications that are one-to-many or client-server in design. Applications that would require end-user customers to deliver content as quickly as they receive it are limited by asymmetric bandwidth. Asymmetric network architecture supports the cable and telephone companies' triple-play business plans, which focus on end users as receivers, rather than transmitters, of information. This is almost certainly consistent with current demands of most customers. If customer demand were to move toward applications and services requiring more symmetric downloading and uploading capability — perhaps as a result of heightened popularity for interactive games or peer-to-peer distribution of videos and other files — the current asymmetric architecture might constrain the growth of these applications, but it also might create market forces for entry of a third broadband provider with a more symmetric network or for the incumbents to modify their networks to meet the new demand.

A network provider can make other decisions about its last mile network that will affect the bandwidth available to end users. Whether its last mile architecture is all fiber, fiber and copper, or fiber and coaxial cable, it can choose to deploy electronics that determine the bandwidth capacity of the line into the end-user premise. In addition, it can partition the bandwidth capacity of the line into the end-user premise, reserving some portion or portions of the total bandwidth for specific applications. For example, a provider might reserve a portion of the bandwidth for its own applications or for those of an independent applications provider that pays for priority access to the end user.

Each network provider can make its own decisions about these technical parameters, subject to market constraints (though currently not subject to regulatory constraints). For example, Verizon's fiber-to-the-home last mile architecture could potentially provide almost limitless bandwidth if all the fiber strands were "lit," but demand cannot justify the deployment of the electronics needed for such unlimited capacity. As currently configured, Verizon's Fios offering lights just three of the many fiber strands in the optical fiber that comes to the customer premise. These lit strands are called lasers. One of these lasers is reserved for the "broadcast" downloading of all the video channels offered in Verizon's video service. The second laser brings 100 megabits per second (mbps) of bandwidth into the customer premise for downloading packets of all other "incoming" traffic — incoming web pages, e-mails, and other data received as part of internet access service, incoming voice packets, incoming video-on-demand programming,[16] and incoming special services. The third laser is used for uploading packets of all outgoing traffic (associated with these internet access, voice, video-on-demand, and special virtual private network services).

A Verizon end user does not purchase or use the full 100 mbps of bandwidth in the download laser, though that much bandwidth comes into her premise on that laser. Nor does she purchase or use all of the bandwidth in the upload laser. Rather, she purchases specific services that use up to a ceiling level of bandwidth in those lasers. For example, an end user can choose internet access service options with 10, 20, or 30 mbps of downstream bandwidth and with 2 or 5 mbps of upstream bandwidth.[17] The end user also can purchase Verizon's video-on-demand service and/or Verizon's voice service, and have these delivered over the download and upload lasers; or she can purchase an independent applications providers' video-on-demand and/or voice services and have these delivered over the download and upload laser. The remainder of the bandwidth on those lasers — typically 70 mbps of bandwidth on the download laser and substantial bandwidth on the upload laser —are available for special services for that end user. But the end user cannot directly purchase that bandwidth for its own use; rather an independent applications provider that would like to offer a special service to the end user would purchase the bandwidth from Verizon and then recover its costs in its charges to the end user for the service provided over that bandwidth. More specifically, Verizon would require the independent applications provider to purchase an end-to-end connection from its location to the end-user premise. Verizon has characterized these as virtual private network (VPN)-like services that allow the independent applications provider to avoid congestion in the public internet and provide a guaranteed quality of service.

This end-to-end connection is shown in Figure 1 by the link that goes directly from the independent applications provider to the service router in the broadband network provider's proprietary IP cloud and then through the broadband last mile to the end user. Although this VPN-like service shares the download and upload lasers with internet access and other services, Verizon is able to manage the traffic on the lasers to assure that the quality of service for that special service is not degraded by other traffic on those lasers (and, as will be discussed below, the special service packets are accorded priority as they are transmitted across servers and links within Verizon's proprietary IP network). Special services that might be provided over such a VPN-like link could range from home monitoring of medical patients by physicians to high definition video streaming. With respect to the services that Verizon itself offers over the three lasers (currently, video, voice, video-on-demand, and internet access services), customers can choose to purchase one or more of the services; discounts are provided for purchasing multiple services. But a Fios customer who does not receive any special VPN-like services could not use the extra bandwidth on the download and upload lasers for its own purposes, such as extra bandwidth for internet access. Nor could an end user who does not purchase Verizon's video service purchase the bandwidth on the video laser for other purposes.

AT and T will be serving most of its Project Lightspeed end-user customers with its last mile network comprised of fiber to the node and a DSL link to the customer premise. The link can provide up to 25 mbps of bandwidth into the premise. Of that bandwidth, 19 mbps currently are reserved for AT and T's video service. Of the remaining 6 mbps of bandwidth, the customer can purchase 1.5 mbps service, 3 mbps service, or 6 mbps service for internet access and voice services. AT and T's proprietary ethernet IP network controls the bandwidth to each premise, based on the level of bandwidth the end user has purchased. AT and T's network is capable of reassigning some of the 19 mbps currently reserved for the proprietary video service to other uses. It would be technically possible for a customer to order additional bandwidth on demand — for example, 10 mbps of bandwidth for three hours for a particular application; this capability is called "turbocharging." But AT and T is concerned that last mile congestion that harms service quality could occur if a household attempted to use AT and T's video service and also an internet application requiring more than 6 mbps of bandwidth at the same time. Therefore AT and T does not currently offer a turbocharge (bandwidth-on-demand) service.

According to CableLabs, the industry research consortium that has developed the Data Over Cable Service Interface Specifications (DOCSIS) that define interface standards for cable modems and supporting equipment, typically a few hundred cable end-user subscribers get internet access by sharing a 6 megahertz (MHz) downstream channel (from the cable network to the customer) and one or more upstream channels (from the customer to the cable network).[18] The downstream channel occupies the space of a single television transmission channel in the cable operator's channel lineup and can provide up to 40 mbps of bandwidth. The cable modems that are most widely deployed in cable networks today (which meet DOCSIS 1.0 and 1.1 specifications) allow upstream channels to deliver up to 10 mbps of bandwidth. Cable companies are now deploying cable modems that meet DOCSIS 2.0 standards that allow upstream channels to deliver up to 30 mbps. DOCSIS 3.0, currently in the late stages of development, will allow several downstream and several upstream channels to be bonded together to multiply the bandwidth delivered to each customer.

Comcast has recently announced that it is rolling out a free feature, called "Powerboost," that will give end users a temporary turbocharge, doubling speeds for many downloads.[19] Comcast currently has one service offering of 6 mbps downstream/384 kbps upstream and a second offering of 8 mbps downstream/768 kbps upstream for internet access and voice service. With the Powerboost feature, customers of these offerings would be able to enjoy downstream speed bursts of 12 and 16 mbps, respectively. The remainder of the bandwidth capacity of the hybrid fiber/coaxial cable lines into their premises would continue to be reserved for video channels. Comcast does not guarantee these speeds for its internet access service. Since many customers share a single channel, the actual speed available to an individual customer at any specific point in time will depend on the level of usage by neighboring customers who share the channel.

Broadband Network Providers' Proprietary IP Networks

As shown in Figure 1, independent applications providers can access a broadband network either through the internet backbone or through a direct connection to a service router in the network provider's proprietary IP network.

The broadband network providers are constructing proprietary IP networks that have the intelligence needed to manage overall traffic flow in their networks as well as the flow of traffic to individual end users. Typically, the network providers are deploying these IP networks in each of the metropolitan areas in which they offer service. As shown in Figure 1, traffic will arrive at the service router from a number of sources — from the internet backbone, directly from an independent applications provider, or from the network provider's own video, voice, or data server.[20] The service router, communicating with other portions of the proprietary IP network, is programmed to determine the route the incoming traffic will take to reach the end user and to prioritize traffic in order to determine which packets may be delayed or dropped during periods of congestion. This task includes setting the priority algorithm employed at the router to determine which packets are delayed or dropped when congestion occurs. It is possible that a prioritization algorithm could reserve certain links for particular prioritized packets, such that even if there is bandwidth available on those links for non-prioritized packets, such non-prioritized packets cannot be transmitted over those links. For example, as discussed earlier, Verizon has indicated that, as part of its Fios service, a customer can purchase a path or laser that would be dedicated to delivery of a VPN-type end-to-end service, such as a medical monitoring service. Verizon has not clarified whether some of the links —or, at least, some partitioned portion of the bandwidth in those links — would be entirely reserved for such VPN service (denying any other packets access to that bandwidth) or if the packets associated with the VPN service simply would be given the highest priority at the service router, with the bandwidth in those links used for lower priority packets when there are no VPN packets.

The proprietary IP network also manages and controls the traffic flow through the broadband last mile to the end user. It has the intelligence to, among other things, partition the last mile, make available to the end user only the bandwidth purchased by that end user, slow down traffic that may be moving too fast for the bandwidth capability of the last mile, and prioritize traffic moving onto the last mile.[21] Embedded in the proprietary IP network also is the capability to temporarily turbocharge end user lines to accommodate bandwidth-intensive applications or to allow for flexible partitioning of the last mile so that bandwidth that is normally partitioned and reserved for a specific use might be made available to accommodate a different use.[22]

These proprietary IP networks consist of a physical (transmission) network layer, a logical layer (usually the transmission control protocol/internet protocol suite of protocols (often referred to as TCP/IP), which itself consists of several layers), an applications layer, and a content layer.[23] It is technically possible for an independent applications provider to gain access to a broadband network at various layers, with that provider providing more or less of its own intelligence depending on the layer at which access occurs. Some independent applications providers have alleged that they have been denied access at a layer that would allow them to use their own IP capabilities to differentiate their products from those of the network provider — for example, to offer unique filtering services that might be desired by families who want more restrictive program filters than those offered by the broadband network or to offer robustly secure internet service with special intrusion detectors capable of stopping DOS attacks that the broadband network provider might not offer.[24] These independent providers claim that, instead, they were given access only in a fashion that would allow them to resell Verizon's Fios product offering. Verizon responds that, with the advent of its video service, it had to direct data associated with the video service separately from data associated with information service provider (ISP) service, and this required a change in the way it provided independent ISPs access to its network.[25] The independent ISPs question whether such a change was really needed to upgrade the network or whether it was a strategic decision that undermines their ability to differentiate their applications from Verizon's.

The Internet Backbone

As explained in footnote 4, the internet largely consists of a number of routers with links leading into and out of those routers. Traditionally, traffic has traversed the internet based on a best effort standard in which packets are not prioritized, although routers do need some basis for determining which packets to delay or to drop during periods of congestion.

The two largest telephone companies, AT & T and Verizon, are among the largest providers of internet backbone facilities. Other companies with substantial internet backbone companies include Sprint-Nextel, Level3, and Qwest. The cable companies have very limited internet backbone facilities.

There is some question as to whether the telephone companies consider their internet backbone facilities to be part of their proprietary networks, in which they would program their routers to prioritize packets. Referring again to Figure 1, AT & T and Verizon have made it clear that they intend to program the service routers in their proprietary IP networks to prioritize incoming packets. But AT & T and Verizon own routers and links in the internet backbone, as well, and they have not clarified whether they intend to program those routers to prioritize packets too.

This distinction could have important public policy implications. If the telephone companies were to, in effect, extend their proprietary IP network into the traditional internet backbone by programming prioritization into their routers in the internet backbone, then if and when congestion occurred at any of the telephone companies' internet backbone routers, non-prioritized packets might be delayed or even dropped. If it were possible and relatively inexpensive to identify the telephone companies as the source of these delayed or dropped packets and to route traffic away from those prioritizing routers, non-telephone company internet backbone providers might be able to expand their capacity and attract customers and traffic away from the telephone companies' internet backbone facilities. But these efforts, if doable, would not be costless. Some independent applications providers — and their customers — could well be harmed by degraded service. (It also is possible that the cable companies, to the extent they use the public internet to offer their applications, could be harmed, since they do not have their own internet backbone facilities.)

If, on the other hand, the telephone companies only prioritized packets once those packets were at the service routers in the telephone companies' private IP networks, then such prioritization is unlikely to degrade the quality of service within the internet backbone itself. Even if the telephone companies chose to focus their investments on their proprietary networks and chose not to upgrade their internet backbone facilities, as long as there continued to be demand for transport over the public internet, then the many non-telephone company providers of internet backbone facilities would still have the incentive to expand and upgrade their internet backbone facilities.

Parameters Available to Broadband Network Providers that Can Affect End Users' and Independent Applications Providers' Access to These Networks

Broadband network providers have many parameters within their control that can affect end users' and independent applications providers' access to these networks. These include:

- the choice of the last-mile network architecture: fiber to the home, fiber to the curb or node (hybrid fiber-DSL), or hybrid fiber-coaxial cable.
- the choice between "broadband" last-mile architecture that transmits all the multiple channel video signals all the way to the end-user premise and "call-up" architecture that only transmits to the end-user premise the particular video channel selected by the customer at the set-top box.
- the choice between more or less symmetry in the network, in terms of bandwidth capacity for an end user to download (receive) a file transmitted over the network vs. bandwidth capacity to upload (send) a file transmitted over the network.
- the choice between deploying a network with very great bandwidth and limited ability to manage traffic congestion (presumably because the bandwidth will be sufficient to minimize congestion) and deploying a network with less bandwidth, but a greater need and ability to manage traffic through prioritization and other capabilities.
- the choice of electronics deployed in the network to turn potential bandwidth capacity into actual available capacity. This includes choices about both the optical fiber strands (lasers) to light and the capability and number of cable or DSL modems deployed.
- the choice of whether and how to partition the bandwidth in both the last-mile connections and the links in the proprietary IP network. This includes choices about how much bandwidth to partition for particular prioritized uses, what those prioritized uses are, and whether the partitioning is flexible (i.e., able to be changed when actual usage patterns result in unused bandwidth in a partitioned portion of a link reserved for prioritized packets while congestion is creating delay or other latency problem for the non-prioritized packets using the non-partitioned portion of the link).
- the choice of where and how to prioritize packets. This includes choices about whether to prioritize packets only within the broadband network provider's proprietary broadband network or also at routers in the internet backbone; what basis to use for delaying individual packets when there is congestion; and what basis to use for dropping individual packets when a router's memory is full.
- the choice of what to include in specific service offerings and the prices for those service offerings. This involves both service offerings to end users and service offerings to independent applications providers. It includes choices about how many services to bundle together, whether to make services available only as part of a bundle, whether to offer different bandwidth options, what those bandwidth options are, what the prices are for each service offering and option, whether to charge end users or independent applications providers for bandwidth, the price to end users

and/or independent applications providers for bandwidth relative to the price of the network provider's end-user service offerings, any usage restrictions in the service offerings for either end users or independent applications providers, the tier or level in the IP network at which independent applications providers gain access, the extent to which an independent applications provider can employ its own IP capability as well as the IP capability in the broadband network to offer service, and the quality of service guarantees in service offerings for independent applications providers.

These parameters are interactive. For example, the greater the bandwidth capacity of the network, the less the need for partitioning or prioritizing traffic or for imposing any use restrictions.

The specific choices that broadband network providers make about these parameters will be driven by several forces — the relative costs of the network architecture options, the actual and perceived demand for (and price sensitivity of) the various service offerings, the actual and potential competition for the provision of both broadband network services and applications, statutory or regulatory constraints (if any), and, to the extent the network providers enjoy some degree of market power, strategic considerations.

How Might Strategic Broadband Network Provider Behavior Harm Consumers?

Generally, broadband network providers will not want to take actions that restrict the availability or quality of applications that end users can obtain over their networks. Such restrictive behavior would reduce overall demand for the broadband network and also increase incentives for competitive entry. At the same time, to the extent that the broadband network providers seek to maximize their revenues for what they perceive as the killer broadband applications — voice and video service today, perhaps interactive games or other applications in the future — they will have an incentive to build, operate, and manage their broadband network in a fashion that favors their own applications over competitors' applications. With only limited alternatives to the cable and telephone broadband duopoly for the foreseeable future, and with the cable and telephone companies both pursuing largely the same business plan, the broadband providers might have both the incentive and the ability to exploit their control over access to end users to restrict competition (and the innovation it might bring) and harm consumers. This strategic behavior could occur in several ways.

Given its control over the bandwidth capacity of its network and the partitioning of that bandwidth, if the network provider were to reserve a substantial portion of the bandwidth (in the last-mile network as well as in the links of its proprietary IP network) for its own latency-sensitive services, in order to assure a particular quality of service for those services, that might leave too little bandwidth available for independent applications to assure an equal quality of service for those independent applications. Or there might not be sufficient bandwidth available for multiple independent applications providers to simultaneously serve a single premise or for the provision of certain bandwidth-intensive applications from independent providers.

For example, streaming a high definition television (HDTV) channel currently requires approximately 20 mbps of bandwidth, though advances in compression technology are likely to reduce the required bandwidth to 10 mbps in the near future. If a broadband network provider reserved most of the bandwidth into a customer's premise for its own video service, leaving at most 6 mbps available for the internet access needed to receive independent applications, competitive provision of HDTV would be curtailed. This might, or might not, represent anticompetitive strategic behavior. If the broadband provider has deployed fiber to the home, bringing 100 mbps to the end user premise, and then limits its internet access service offering to 6 mbps, while providing multiple HDTV channels over its own video service, this might suggest an anticompetitive strategic partitioning decision. On the other hand, if the broadband provider has deployed a fiber-DSL hybrid network that only brings 25 mbps to the premise, and its own partitioned video service does not offer HDTV capability (or only limited HDTV capability), then if its internet access service is only 6 mbps, this might reflect network limitations rather than, or as well as, strategic behavior. In either situation, however, if it were technically and economically feasible to partition the bandwidth flexibly, so a customer could use 10 mbps of the bandwidth coming to its premise for either the broadband network's HDTV service offering or an independent application provider's HDTV service offering, the consumer is likely to enjoy greater choice in applications. But even this result is not unambiguous. In a household with multiple high definition television sets, if that household were simultaneously streaming multiple HDTV programs from both the network provider and from an independent provider, flexible partitioning might allow all the programs to be viewed, but not be able to ensure the maintenance of HDTV quality for either program.[26]

Given that the broadband network providers are providers of both end user services and input (network access) services required by their independent applications competitors, they may have the opportunity to set prices for their network access and applications services in a strategic fashion. Consider, for example, the prices that Verizon currently charges its Fios customers for internet access. Verizon offers three options: up to 5 mbps download speed and 2 mbps upstream speed for $34.95, up to 15 mbps download and 2 mbps upload for $49.95, and up to 30 mbps download and 5 mbps upload for $179.95.[27] There are several possible explanations for the huge jump in price for the 30 mbps service. One explanation might be that Verizon would have to incur substantial costs increasing the capacity of its last mile network (and perhaps its proprietary IP network) to handle those bandwidth-intensive applications that would require 30 mbps of download bandwidth. In this case, the high price would accurately reflect actual underlying costs. A second possible explanation might be that the customers with such substantial bandwidth needs tend to be insensitive to price and thus will pay very high prices, or are part of a category of customers (such as business customers) who have traditionally been charged higher rates. Then, the high price would represent a way to perform efficient price discrimination to recover fixed network costs. A third possible explanation might be that Verizon faces potentially strong competition from independent applications providers for the provision of bandwidth-intensive applications, such as HDTV, and by pricing the 30 mbps of internet access service needed for those services at $179.95, while charging a lower prices for its own HDTV or other bandwidth-intensive applications, it could practice strategic behavior that places its competitors in an anticompetitive price squeeze.

More generally, access to the bandwidth provided by broadband networks is a necessary input into the provision of broadband applications. The broadband providers have announced that they seek compensation for such broadband access in two ways — in charges to end users for specific quantities of bandwidth access and in charges to independent applications providers for prioritized access to the broadband network (intended to guarantee service quality, typically comparable to the quality the broadband network providers provide themselves for their own applications). For a particular application, if the difference between the price that the network provider charges end users for its own application service and the "imputed" cost of access to the broadband network that it would have to pay if it were an independent applications provider offering that application (that is, the charges that the network provider imposes on end users and independent applications providers for the bandwidth needed to offer that application) is less than the non-bandwidth-related costs for the network provider to offer that application, then the network provider is placing the independent applications provider in a price squeeze because even if that independent provider were just as efficient as the network provider it would be placed at a competitive disadvantage simply due to the network provider's pricing decisions. This would allow the network provider to succeed in the applications market despite being a less efficient provider.

Proposals for Government Policy on Access to Broadband Networks

There have been a number of proposals for government policy on access to broadband networks and how best to addressing broadband network providers' strategic behavior.

The FCC Broadband Policy Statement

H.R. 5252, which has been passed by the full House, explicitly authorizes the FCC (at Sec. 201) to enforce the broadband policy statement, and the principles incorporated therein, that the Commission adopted as general principles on August 5, 2005. These principles are:

- consumers are entitled to access the lawful internet content of their choice.
- consumers are entitled to run applications and use services of their choice, subject to the needs of law enforcement.
- consumers are entitled to connect their choice of legal devices that do not harm the network.
- consumers are entitled to competition among network providers, application and service providers, and content providers.

H.R. 5252 explicitly prohibits the FCC from adopting or implementing rules or regulations regarding enforcement of the broadband policy statement, except to adopt procedures for the adjudication of complaints.

These principles are quite general and susceptible to alternative interpretations. They would prohibit a broadband network provider from entirely blocking a particular application, such as a competitor's VoIP service. They do not explicitly prohibit a broadband network provider from prioritizing packets or reserving significant portions of bandwidth for its own applications or for the applications of a preferred independent provider, even if such behavior harmed the quality of service of one or more independent applications providers or effectively precluded independent applications providers from the market. Nor do they explicitly prohibit a broadband network provider from setting charges for network access in a fashion that would place independent applications providers in an anticompetitive price squeeze. Arguably, the fourth principle — consumer entitlement to competition among application and service providers — could be the basis for a complaint against such behavior. No standards are provided, however, for determining what level of competition a consumer is entitled to. For example, the FCC might view the duopoly provision of applications, by the telephone and cable companies, sufficient competition to meet this principle, even if independent applications providers were harmed, or even excluded from the market, by the behavior.

Internet Consumer Bill of Rights

The Internet Consumer Bill of Rights, incorporated in section 903 of H.R. 5252 as amended by the Senate Commerce Committee (originally introduced as S. 2686), requires each internet service provider to allow each subscriber to:

- access and post any lawful content of that subscriber's choosing;
- access any web page of that subscriber's choosing;
- access and run any voice application, software, or service of that subscriber's choosing;
- access and run any video application, software, or service of that subscriber's choosing;
- access and run any search engine of that subscriber's choosing;
- access and run any other application, software, or service of that subscriber's choosing;
- connect any legal device of that subscriber's choosing to the internet access equipment of that subscriber, if such device does not harm the network of the internet service provider; and
- receive clear and conspicuous information, in plain language, about the estimated speeds, capabilities, limitations, and pricing of any internet service offered to the public.

This bill of rights addresses only consumer access to those applications, services, or devices that independent providers are able to offer. If, as a result of a network provider's prioritization, partitioning, and/or pricing decisions, an independent applications provider is not able to offer an application — or can only offer an application that is inferior in quality or higher in price than the competing service offered by the broadband network provider — the independent applications provider has no recourse. The consumer has not been denied access; the provider has been denied access. Arguably, under the Internet Consumer Bill of Rights, a consumer who, because of partitioning, is not allowed to purchase sufficient bandwidth to support the bandwidth-intensive offering of an independent applications provider, but could purchase from her broadband network provider a competing bandwidth-intensive service offering, could bring a complaint that she has been denied access to an application of her choosing. But, unlike the FCC principles, the Internet Consumer Bill of Rights does not include a consumer right to competitive options, and thus cannot address anticompetitive behavior or unfair practices on the part of a broadband network provider.

Network Neutrality

Network neutrality has been a buzz-phrase in the on-going debates, though there is not a single, agreed-upon definition of network neutrality. Indeed, there continue to be questions about what constitutes "neutrality" and to which "networks" such neutrality would apply. For most proponents, network neutrality requires all packets to be treated the same way or, at the least, all packets providing a particular application (such as voice or video) to be treated the same way. Some proponents who would allow for prioritization among applications nonetheless would not allow broadband network providers to charge independent applications providers for such prioritization. Network neutrality proposals include provisions relating to consumer access similar to those found in the FCC principles and Internet Bill of Rights, but in addition have provisions relating to nondiscriminatory or neutral access to broadband networks or the internet by independent applications providers. Three such proposals have been incorporated in legislation introduced in the 109th Congress — the Network Neutrality Act of 2006 (H.R. 5273), the Internet Non-Discrimination Act of 2006 (S. 2360), and the Internet Freedom Preservation Act (S. 2917).

Under H.R. 5273, each broadband network provider has the duty to:

- offer, upon reasonable request to any person, a broadband service for use by such person to offer or access unaffiliated content, applications, and services;
- not discriminate in favor of itself in the allocation, use, or quality of broadband services or interconnection with other broadband networks;
- offer a service such that content, applications, or service providers can offer unaffiliated content, applications, or services in a manner that is at least equal in the speed and quality of service that the operator's content, applications, or service is accessed and offered, and without interference or surcharges on the basis of such content, applications, or services;
- if the broadband network provider prioritizes or offers enhanced quality of service to data of a particular type, prioritize or offer enhanced quality of service to all data of that type (regardless of the origin of such data) without imposing a surcharge or other consideration for such prioritization or quality of service; and
- not install network features, functions, or capabilities that thwart or frustrate compliance with the requirements of objectives of this section.

Under S. 2360, a network operator shall:

- not discriminate in favor of itself or any other person, including any affiliate or company with which such operator has a business relationship, in (A) allocating bandwidth; and (B) transmitting content or applications or services to or from a subscriber in the provision of a communications; ! not assess a charge to any application or service provider not on the network of such operator for the delivery of traffic to any subscriber to the network of such operator;

- offer communications such that a subscriber can access, and a content provider can offer, unaffiliated content or applications or services in the same manner that content of the network operator is accessed and offered, without interference or surcharges;
- treat all data traveling over or on communications in a nondiscriminatory way;
- offer just, reasonable, and nondiscriminatory rates, terms, and conditions on the offering or provision of any service by another person using the transmission component of communications; and
- provide nondiscriminatory access and service to each subscriber.

Under S. 2917, each broadband service provider shall:

- enable any content application, or service made available via the internet to be offered, provided, or posted on a basis that (A) is reasonable and nondiscriminatory, including with respect to quality of service, access, speed, and bandwidth; (B) is at least equivalent to the access, speed, quality of service, and bandwidth that such broadband service provider offers to affiliated content, applications, or services made available via the public Internet into the network of such broadband service provider; and (C) does not impose a charge on the basis of the type of content, applications, or services made available via the internet into the network of such broadband service provider;
- only prioritize content, applications, or services accessed by a user that is made available via the internet within the network of such broadband service provider based on the type of content, applications, or services and the level of service purchased by the user, without charge for such prioritization; and ! not install or utilize network features, functions, or capabilities that impede or hinder compliance with this section.

These three network neutrality proposals have similarities and differences. S. 2360 appears to be the most restrictive. It would prohibit a network provider from prioritizing traffic. Thus, the service router could not be programmed to favor the packets of latency-sensitive applications, such as voice or video service. The proposal also would prohibit a network provider from charging an independent applications provider for the delivery of traffic. All such charges would have to be imposed directly on end users. It also appears to prohibit a network provider from reserving bandwidth for its own, or any other provider's, applications. Referring to Figure 1, an independent applications provider could still choose to purchase a direct connection to the service router in the network provider's proprietary IP network, rather than routing its traffic through the internet backbone. But at that service router, all packets would have to be given the exact same priority, whatever the particular type of service that packet was providing, and whether that packet was carrying bits from the network provider's own application, bits from the application of an independent applications provider that had a direct connection to the service router, or bits from the application of an independent service provider that had transmitted those bits through the internet backbone. In addition, neither the links transmitting the packets from the service router to the broadband providers end office/headend nor the last mile connection to the end-user premise could be partitioned in a fashion that would favor the broadband network's applications, for example, by limiting the amount of bandwidth available for independent applications in a way that would not ensure the same quality of service for independent applications and the broadband provider's applications.

H.R. 5273 is less restrictive. It allows the network provider to prioritize traffic — although all traffic for a given application must be assigned the same priority and there can be no surcharge imposed on the higher priority traffic. Referring to Figure 1, an independent applications provider could still choose to purchase a direct connection to the service router in the network provider's proprietary IP network, rather than routing its traffic through the internet backbone. At that service router, prioritization could occur, to reduce the risk of delay in packets for latency-sensitive applications, but all packets with bits for a particular type of service (for example, video service) would have to be given the exact same priority, whether those packets were carrying bits from the network provider's own video service, bits from the video service of an independent applications provider that had a direct connection to the service router, or bits from the video service of an independent applications provider that had transmitted those bits through the internet backbone. Like S. 2360, under H.R. 5273, neither the links transmitting the packets from the service router to the broadband providers end office/headend nor the last mile connection to the end-user premise could be partitioned in a fashion that would favor the broadband network's applications, for example, by limiting the amount of bandwidth available for independent applications in a way that would not ensure the same quality of service for independent applications and the broadband provider's applications. Also like S. 2360, H.R. 5273 would prohibit a network provider from charging an independent applications provider for the delivery of traffic. All such charges would have to be imposed directly on end users.

The current network configurations and service offerings of Verizon, AT & T, and Comcast would not appear to meet the requirements in these two network neutrality proposals because they all reserve significant bandwidth for their own video services and do not appear to make an equal amount of bandwidth available for other applications providers. Also, to the extent that these broadband networks incorporate partitioning and prioritization, and charge accordingly, they do not conform with all the network neutrality requirements.

But partitioning and prioritization need not have discriminatory or anticompetitive consequences. For example, the current Fios offering, which brings 100 mbps of bandwidth to a customer premise on a download laser, and offers up to 30 mbps of that bandwidth for downloading any service, appears to make available enough bandwidth for independent applications providers to be able to compete with Verizon's applications in the provision of even bandwidth-intensive applications.

It is not clear how the network neutrality conditions would address a price squeeze. On one hand, the prohibition on surcharges for prioritization or for network access would appear to constrain the ability of a network provider to use charges imposed on independent applications providers to create a price squeeze. On the other hand, there does not appear to be any constraint on the price that end users can be charged for internet access, and if the price of the network provider's applications do not cover the imputed cost of such access plus the other costs of offering the service, the network provider could create an anticompetitive price squeeze.

S. 2917 appears to have an additional point of departure that might result in virtually no restrictions on broadband providers. The duties of broadband service providers outlined in sections 12(a)(4) and (5) apply to "service made available via the Internet," without further delineation. But there may be differences of opinion about which services are offered via the internet. Consider, for example, the VPN-like services available as part of Verizon's Fios offering, discussed earlier. Those services would be provided by independent applications providers that would purchase the bandwidth from Verizon in order to have an end-to-end connection with the end user; the independent applications provider who purchases that bandwidth would be compensated by charging the end user for the service provided. This end-to-end connection is shown in Figure 1 by the link that goes directly from the independent applications provider to the service router in the broadband network provider's proprietary IP cloud and then through the broadband last mile to the end user. Arguably, since the traffic transported over this VPN service never touches the public internet, but rather goes from an independent applications provider directly to Verizon's proprietary IP network over a private line connection, it is not "a service made available via the Internet." With that interpretation, services offered over this VPN path would not be subject to the nondiscrimination provisions relating to prioritization, partitioning, and charges in sections 12(a)(4) and (5).

More significantly, as shown in Figure 1, the broadband network providers' video services do not use the public internet; they receive satellite, terrestrial, or broadcast feeds into their video server and then route those signals through the service router in their proprietary IP networks and then through their last mile networks to subscribers. The nondiscriminatory bandwidth and prioritization provisions in section 12(a)(4) and (5) therefore may not apply to the network providers' video services or any other applications that are not delivered through the public internet.

Antitrust and Unfair Methods of Competition Laws

There have been several proposals that anticompetitive broadband network access behavior be addressed through application of existing or modified antitrust and competition laws.

The Internet Freedom and Nondiscrimination Act of 2006 (H.R. 5417), which has been ordered to be reported out of the House Judiciary Committee, would amend the Clayton Act by inserting a new section 28 making it unlawful for any broadband network provider:

- to fail to provide its broadband network services on reasonable and nondiscriminatory terms and conditions such that any person can offer or provide content, applications, or services to or over the network in a manner that is at least equal to the manner in which the provider or its affiliates offer content, applications, and services, free of any surcharge on the basis of the content, application, or service;
- to refuse to interconnect its facilities with the facilities of another provider of broadband network services on reasonable and nondiscriminatory terms and conditions;

- to block, to impair, to discriminate against, or to interfere with the ability of any person to use a broadband network service to access, to use, to send, to receive, or to offer lawful content, applications, or services over the Internet;
- to impose an additional charge to avoid any conduct that is prohibited by this subsection;
- to prohibit a user from attaching or using a device on the provider's network that does not physically damage or materially degrade other users' utilization of the network; or
- to fail to clearly and conspicuously disclose to users, in plain language, accurate information concerning any terms, conditions, or limitations on the broadband network service.

In addition, under H.R. 5417, if a broadband network provider prioritizes or offers enhanced quality of service to data of a particular type, it must prioritize or offer enhanced quality of service to all data of that type (regardless of the origin or ownership of such data) without imposing a surcharge or other consideration for such prioritization or enhanced quality of service.

Since enforcement of the Clayton Act is solely within the jurisdiction of either the Federal Trade Commission (FTC) or the Antitrust Division of the Department of Justice (DOJ), the general thrust of H.R. 5417 is to implement the network neutrality provisions in H.R. 5273, but to move enforcement responsibility from the FCC to the antitrust agencies, which specialize in competition analysis and, in the case of the FTC, consumer protection analysis. The FTC has indicated that it believes it already possesses sufficient authority to adequately address the competition and consumer protection issues of concern to network neutrality proponents,[28] but "any new legislation should clearly preserve the FTC's existing authority over activities currently within its jurisdiction."[29]

H.R. 5417 also would have the effect of creating a private right of action by aggrieved applications providers. Although H.R. 5417 contains no explicit language to that effect, section 4(a) of the Clayton Act[30] authorizes treble-damage suits in appropriate district courts by "any person who shall be injured in his business or property by reason of anything forbidden in the antitrust laws...."

The Digital Age Communications Act of 2005 (S. 2113), which is based in large part on a proposal developed as part of the Digital Age Communications Act (DACA) project of the Progress and Freedom Foundation,[31] would replace the current regulatory framework with a framework based on the Federal Trade Commission Act (15 U.S.C. 41 et seq.). It would prohibit unfair methods of competition, and give the FCC the authority to enforce the law. Specifically:

- it shall be unlawful for any provider of electronic communication service to engage or participate, or attempt to engage or participate, in unfair methods of competition, or unfair or deceptive practices in or affecting electronic communications networks and electronic communications services;
- the FCC may, by rule, define with specificity, the acts or practices that shall constitute unfair methods of competition or unfair or deceptive acts or practices;
- promulgated rules may include such requirements as the FCC determines necessary to prevent any methods, acts, or practices prohibited by this section;

- the FCC shall have no authority to issue rules that declare unlawful an act or practice on the grounds that such act or practice is an unfair method of competition or unfair or deceptive act or practice, except that the FCC may declare an act or practice unlawful if the FCC determines, based on a showing of clear and convincing evidence presented in a rulemaking proceeding that (i) marketplace competition is not sufficient to adequately protect consumer welfare, and (ii) such act or practice (I) causes or is likely to cause substantial injury to consumers, and (II) is not avoidable by consumers themselves and not outweighed by countervailing benefits to consumers or to competition;
- the FCC shall have authority to hear complaints from any party injured by a violation of the prohibitions established and to award damages to such injured party if the FCC determines that a violation has occurred.

Although these provisions lack specificity as to what types of behavior would represent unfair methods of competition or unfair or deceptive practices, and in particular do not provide any specificity about discriminatory practices, they do potentially provide a procedure for independent applications providers to file complaints about the terms, conditions, and prices under which they have access to a broadband provider's network. The effectiveness of such a complaint procedure, however, may depend on the standard used in determining whether a particular act or practice represents a legal violation.

On May 10, 2006, in response to questions after his keynote speech at an American Enterprise Institute conference on Key Issues in Telecommunications Policy, Alfred Kahn, who was a DACA Project Advisory Committee Member, commented on the DACA proposal incorporated in S. 2113.[32] He stated that he agreed with the overall framework, but that he believed the proposed standard for finding a violation was set too high. He suggested the standard for a finding of unfair methods of competition or unfair competitive practices should be based solely on demonstrated harm to consumers. He voiced concern with the methodology frequently employed in antitrust analysis today that assumes that no intervention is needed if it can be shown that each $1 loss in consumer surplus is matched by a $1 gain in producer surplus. He argued that unfair practices laws are intended to protect consumers, even if there might be some negative impact on producer efficiency. He therefore would want unfair methods of competition to be "close to *per se* illegal."

References

[1] See George Ford, Thomas Koutsky, and Lawrence Spiwack, "Competition After Unbundling: Entry, Industry Structure and Convergence," Phoenix Center Policy Paper Number 21, July 2005, available at [http://www.phoenix-center.org/pcpp/ PCPP21 Final.pdf], viewed on May 24, 2006.

[2] According to the most recent FCC report, *High-Speed Services for Internet Access: Status as of June 30, 2005,* April 2006, at Table 6, the vast preponderance of high-speed internet access lines were provided by local telephone companies or cable companies: of the 42.9 million total high-speed line (over 200 kilobits per second (kbps) in at least one direction) in the U.S. in June 2005, 14.3 million were provided by Regional Bell

Operating Companies, 2.3 million were provided by other incumbent telephone companies, and 23.9 million were provided via cable modems. Tables 3 and 6 of that report indicate that of the 38.5 million high-speed lines designed to serve primarily residential end users, 61.0% used cable modems and 37.2% were asymmetric digital subscriber lines (ADSL), and that the vast majority of the latter were provided by local telephone companies.

[3] See, for example, Robert D. Atkinson and Philip J. Weiser, "A 'Third Way' on Network Neutrality," The Information Technology and Innovation Foundation, May 30, 2006, at pp. 7-8, available at [http://www.innovationpolicy.org/pdf/netneutrality.pdf], viewed on May 31, 2006.

[4] Latency is affected by physical distance, the number of "hops" from one internet network to another internet network that must be made to deliver the packets (since there can be congestion at each hand-off point), and voice-to-data conversion. Congestion that delays the transmission of packets can cause several problems. (The following is a distillation of a description for lay readers presented by Ed Felten in [http://www.freedom-to-tinker.com?p=983] and [http://www.freedom-to-tinker.com/?p=986], both viewed on May 24, 2006.) In effect, the internet (or a proprietary IP network) is as a set of routers connected by links. Packets of data get passed from one router to another, via links. A packet is forwarded from router to router, until it arrives at its destination. Typically, each router has several incoming links on which packets arrive, and several outgoing links on which it can send packets. When a packet shows up on an incoming link, the router will figure out on which outgoing link the packet should be forwarded. If that outgoing link is free, the packet can be sent out on it immediately. But if the outgoing link is busy transmitting another packet, the newly arrived packet will have to wait — it will be "buffered" in the router's memory, waiting its turn until the outgoing link is free. Buffering lets the router deal with temporary surges in traffic. The router will be programmed to determine which packets should be delayed and also, when the link is available, which buffered packet should be transmitted. That is, a packet prioritization scheme is devised. This could be a simple, first-in, first-out scheme or a favor-applications-sensitive-to-packet-delay scheme, or a pay-for-priority scheme, or something else. But if packets keep showing up faster than they can be sent out on some outgoing link, the number of buffered packets will grow and grow, and eventually the router will run out of buffer memory. At that point, if one more packet shows up, the router has no choice but to discard a packet. It can discard the newly arriving packet, or it can make room for the new packet by discarding something else. But something has to be discarded. The router will be programmed to determine which packets should be dropped, thus creating a second packet prioritization scheme. Again, this could be a simple, first-in, first-out scheme or a favor-applications-sensitive-to-dropped-packets scheme, or a pay-for-priority scheme, or something else. Dropped packets can be retransmitted, but for those applications, such as voice, that require the packets to arrive and be reassembled within a short period of time, such packet recovery might not occur in the timely fashion needed to retain service quality. With such congestion, at least two problems may occur. One problem is dropped packets. Some applications are more sensitive than others to dropped packets. A second problem is "jitter" caused by the delay of certain packets. Internet traffic is usually "bursty," with periods of relatively low activity punctuated by occasional bursts of

packets. (For example, browsing the Web generates little or no traffic while reading the page, but a burst of traffic when the browser needs to fetch a new page.) Even if the router is programmed to minimize delay by only delaying low-priority packets when congestion absolutely requires such delay, if the high-priority traffic is bursty, then low-priority traffic will usually move through the network with little delay, but will experience noticeable delay whenever there is a burst of high-priority traffic. This on-again, off-again delay is called jitter. Jitter has no affect when downloading a big file, for which one's concern is the average packet arrival rate rather than arrival time of a particular packet. But the quality of applications like voice conferencing or VoIP — which rely on steady streaming of interactive, realtime communication — can suffer a lot if there is jitter.

[5] For simplification of exposition, this chapter will refer to those applications that are sensitive to dropped packets or to jitter or any other congestion-related transmission problem as "latency-sensitive" applications.

[6] "P2P in 2005," Presentation by Andrew Parker, co-founder and chief technology officer, CacheLogic, available at [http://www.cachelogic.com/research/ 2005_ slide01.php], viewed on June 9, 2006.

[7] See Julie Bosman and Tom Zeller Jr., "Warner Bros. To Sell Movies and TV Shows on Internet," *New York Times*, May 9, 2006, at p. C3.

[8] See the discussion below in the section entitled "What Do Wireline Broadband Networks Look Like?"

[9] Customer churn is the number of customers who discontinue purchasing service from a provider within a period of time. It is typically measured as the churn rate — the percentage of customers who discontinue service during a period of time.

[10] In the Matter of Inquiry Concerning High-Speed Access to the Internet Over Cable and Other Facilities; Internet Over Cable Declaratory Ruling; Appropriate Regulatory Treatment for Broadband Access to the Internet Over Cable Facilities, 17 FCC Rcd. 4798, 4799 (March 15, 2002). In this decision, the FCC concluded that the telecommunications functionality in cable modem service is integral to the service, and not transparent to the consumer, and therefore cable modem service should be treated as a pure information service and not subject to the requirements imposed on telecommunications services under Title II of the Communications Act (47 U.S.C. 151 et seq.).

[11] National Cable and Telecommunications Association v. Brand X Internet Services, 125 S.Ct. 2688 (2005).

[12] *In the Matter of Appropriate Framework for Broadband Access to the Internet Over Wireline Facilities*, Report and Order and Notice of Proposed Rulemaking, 20 FCC Rcd. 14853 (September 23, 2005). The FCC used the same argument to support its DSL decision as it used in its cable modem decision. Not surprisingly, the United Power Line Council has petitioned the FCC to issue a declaratory ruling that broadband over power lines is an information service akin to cable modem and DSL service. (*In the Matter of the Petition of the United Power Line Council for a Declaratory Ruling Regarding the Classification of Broadband over Power Line Internet Access Service as an Information Service*, Petition for Declaratory Ruling, filed December 23, 2005, available at [http://www.uplc.utc.org/file_depot/0-10000000/0-10000/7966/conman/ Petition+for+De claratory+Ruling.pdf]

[13] For example, at the November 10, 2005, panel of academics at the "Peripheral Visionaries' VoIP Communications Policy Summit" conference sponsored by pulver.com, Susan Crawford of Cardozo Law School, argued that it is unlikely to be possible to craft legislative language that broadband network providers could not get around in the name of network management, and that instead the public policy focus should be on the "real problem" of lack of competitive alternatives to the cable and telephone networks.

[14] AT and T has indicated that, as it builds out its Lightspeed network, it may serve those customers in new housing developments under construction by deploying optical fiber all the way to their premises. These may represent as many as 5% of AT and T customers.

[15] The actual amount of bandwidth provided by fiber, fiber/copper, or HFC will depend on a number of factors, including how much of the optical fiber is "lit" by electronics and the type of modems used.

[16] Verizon's video-on-demand service is provided separately from the rest of its multichannel video service offering because video-on-demand requires communication between the end user and the video-on-demand server in Verizon's proprietary IP network to call up the specific program from the thousands of programs available, whereas the multi-channel video service simply "broadcasts" all the channels in the service to the end-user premise, with no need for prior communication from the end user.

[17] The pricing of these options are discussed below in the section entitled How Might Strategic Broadband Network Provider Behavior Harm Consumers?

[18] See "DOCSIS Project Primer," a document prepared by the CableLabs consortium and last updated September 30, 2005, available at [http://www.cablemodem.com/primer/], viewed on June 6, 2006.

[19] See, for example, Dionne Searcey and Sarmad Ali, "Comcast to Speed Downloads to Some Movie, Music Files," *The Wall Street Journal*, June 1, 2006, at p. D3.

[20] For ease of presentation, the schematic diagram in figure 1 does not attempt to show every link into a broadband network. The voice, data, and video servers are each shown in the diagram with only a link to the service router, but each of these applications servers will also have other links to the sources of voice, data, and video information transmitted. For example, the video server will be connected (by satellite, terrestrial, and broadcast connections) to the sources of the video programming provided by the network provider.Similarly, since lots of voice traffic continues to flow over the traditional public switched telephone network (PSTN), there will continue to be traffic that comes from the PSTN directly into the telephone company's end office in analog format, but that traffic may then be digitized and routed by the telephone company's proprietary IP network.

[21] Prioritization may play an important network management role as peer-to-peer (P2P) applications become more common. One characteristic of P2P networking is to distribute files as broadly through the network as possible, rather than maintaining files at central servers, so that files can be obtained quickly and with less risk from failure at a single mode. Many P2P applications are constructed to identify all the places where a particular file is located, identify which of those places are served by significant uploading bandwidth, and then obtain the file from those locations. These applications

tend to place a burden on last mile networks and proprietary IP networks with large upload capacity. But increasing bandwidth capacity at those locations where P2P applications are creating congestion only creates the feedback effect of making those locations even more attractive for P2P applications. One solution that has been proposed by broadband network providers is for the network manager to be able to employ prioritization in a fashion that slows down P2P packets in order to allow other traffic to flow at these P2P-created congestion points and to discourage the cycle of increases in upload bandwidth intended to reduce congestion automatically attracting additional P2P traffic that creates new congestion. This would help broadband providers manage their networks. But it might diminish efficiencies of P2P networking. Such prioritization is inconsistent with one of the fundamental principles articulated by most proponents of "network neutrality" — that all packets should be treated the same, or that, at the least, all packets associated with the same type of application should be treated the same.

[22] Not every proprietary IP network will necessarily have these capabilities, but announcements by the various broadband network providers suggest that such capabilities are technically possible.

[23] See, for example, Richard S. Whitt, "A Horizontal Leap Forward: Formulating a New Communications Public Policy Framework Based on the Network Layers Model," 56 *Federal Communications Law Journal* 587 (2004).

[24] See Louis Trager, "ISPs Accuse Verizon of Double-Cross on FiOS Wholesaling," *Communications Daily*, November 3, 2005.

[25] Untitled article, *Communications Daily*, November 8, 2005, at p. 8.

[26] But this does highlight how partitioning places constraints on the end user's ability to receive multiple HDTV programs simultaneously from one or more independent applications providers, even if that end user has purchased a bundled service offering with total bandwidth capable of providing multiple HDTV programs simultaneously.

[27] These were the service and price options listed on Verizon's website, on June 22, 2006, for those residential households located in Fairfax County, VA for whom Verizon's Fios service was available.

[28] Behavior that is unfairly discriminatory or anticompetitive is reachable via sections 1 or 2 of the Sherman Act (15 U.S.C. 1and 2, which prohibit, respectively, contracts or conspiracies in restraint of trade and monopolization), or as a violation of section 5 of the FTC Act (15 U.S.C. 45(a), which proscribes "unfair methods of competition" or "unfair or deceptive acts or practices"). The Sherman Act is enforceable by either the FTC or the DOJ; the FTC Act is enforceable only by the FTC..

[29] "FTC Jurisdiction Over Broadband Internet Access Services," Testimony of FTC Commissioner William E. Kovacic before the Senate Judiciary Committee, Hearing on Reconsidering Our Communications Laws: Ensuring Competition and Innovation, June 14, 2006, available at [http://judiciary.senate.gov/print_testimony.cfm?id=1937 and wit_id=5415], viewed on August 31, 2006.

[30] 15 U.S.C. 15(a).

[31] Digital Age Communications Act: Proposal of the Regulatory Framework Working Group of the Digital Age Communications Act Project of the Progress and Freedom Foundation, June 5, 2005, available at [http://www.pff.org/issues-pubs/other/ 050617 regframework.pdf], viewed on June 21, 2006.

[32] Alfred Kahn, keynote address and response to questions, American Enterprise Institute conference on Key Issues in Telecommunications Policy, Washington, DC, May 10, 2006.

In: Computer Networks, Policies and Applications
Editors: C. M. Berger and A. D. Streusse
ISBN: 978-1-61209-088-7

Chapter 2

BROADBAND INTERNET ACCESS AND THE DIGITAL DIVIDE: FEDERAL ASSISTANCE PROGRAMS*

Lennard G. Kruger and Angele A. Gilroy

ABSTRACT

The "digital divide" is a term that has been used to characterize a gap between "information haves and have-nots," or in other words, between those Americans who use or have access to telecommunications technologies (e.g., telephones, computers, the Internet) and those who do not. One important subset of the digital divide debate concerns high-speed Internet access, also known as *broadband*. Broadband is provided by a series of technologies (e.g. cable, telephone wire, fiber, satellite, wireless) that give users the ability to send and receive data at volumes and speeds far greater than current "dial-up" Internet access over traditional telephone lines.

Broadband technologies are currently being deployed primarily by the private sector throughout the United States. While the numbers of new broadband subscribers continue to grow, studies conducted by the Federal Communications Commission (FCC), the Department of Commerce (DOC), and the Department of Agriculture (USDA) suggest that the rate of broadband deployment in urban and high income areas may be outpacing deployment in rural and low-income areas.

Some policymakers, believing that disparities in broadband access across American society could have adverse economic and social consequences on those left behind, assert that the federal government should play a more active role to avoid a "digital divide" in broadband access. One approach is for the federal government to provide financial assistance to support broadband deployment in underserved areas. Others, however, believe that federal assistance for broadband deployment is not appropriate. Some opponents question the reality of the "digital divide," and argue that federal intervention in the broadband marketplace would be premature and, in some cases, counterproductive.

Legislation introduced (but not enacted) in the 109th Congress sought to provide federal financial assistance for broadband deployment in the form of grants, loans,

---

* This is an edited, reformatted and augmented edition of a United States Congressional Research Service publication, Report RL33496, dated August 31, 2006.

Excerpted from CRS Report RL30719, dated February 7, 2007.

subsidies, and tax credits. Many of these legislative proposals are likely to be reintroduced into the 110th Congress. Of particular note is the possible reauthorization of the Rural Utilities Service (RUS) broadband program, which is expected to be considered as part of the 2007 farm bill. Legislation to reform universal service – which could have a significant impact on the amount of financial assistance available for broadband deployment in rural and underserved areas – has been introduced into the 110th Congress (H.R. 42, S. 101).

In assessing such legislation, several policy issues arise. For example, is the current status of broadband deployment data an adequate basis on which to base policy decisions? Is federal assistance premature, or do the risks of delaying assistance to underserved areas outweigh the benefits of avoiding federal intervention in the marketplace? And finally, if one assumes that governmental action is necessary to spur broadband deployment in underserved areas, which specific approaches, either separately or in combination, are likely to be most effective?

BACKGROUND

The "digital divide" is a term used to describe a perceived gap between perceived "information haves and have-nots," or in other words, between those Americans who use or have access to telecommunications technologies (e.g., telephones, computers, the Internet) and those who do not.[1] Whether or not individuals or communities fall into the "information haves" category depends on a number of factors, ranging from the presence of computers in the home, to training and education, to the availability of affordable Internet access. A series of reports issued by the Department of Commerce[2] (DOC) during the Clinton Administration argued that a "digital divide" exists, with many rural citizens, certain minority groups, and low-income Americans tending to have less access to telecommunications technology than other Americans.[3]

In February 2002, the Bush Administration's Department of Commerce released its first survey report on Internet use, entitled *A Nation Online: How Americans Are Expanding Their Use of the Internet.*[4] While acknowledging a disparity in usage between "information haves and have nots," the report focused on the increasing rates of Internet usage among traditionally underserved groups:

> In every income bracket, at every level of education, in every age group, for people of every race and among people of Hispanic origin, among both men and women, many more people use computers and the Internet now than did so in the recent past. Some people are still more likely to be Internet users than others. Individuals living in low-income households or having little education, still trail the national average. However, broad measures of Internet use in the United States suggest that over time Internet use has become more equitable.[5]

A Nation Online: Entering the Broadband Age, published in September 2004, is the sixth Department of Commerce report examining the use of computers, the Internet, and other information technology. For the first time, the DOC report focuses on broadband, also known as high-speed Internet access. Broadband is provided by a series of technologies (e.g. cable, telephone wire, satellite, wireless) that give users the ability to send and receive data at volumes and speeds far greater than current "dial-up" Internet access over traditional telephone lines.[6] The DOC report found that the proportion of U.S. households with broadband connections grew from 9.1% in September 2001 to 19.9% in October 2003.[7]

According to the latest FCC data on the deployment of high-speed Internet connections (released January 2007), as of June 30, 2006 there were 64.6 million high speed lines connecting homes and businesses to the Internet in the United States, a growth rate of 26% during the first half of 2006. Of the 64.6 million high speed lines reported by the FCC, 50.3 million serve residential users.[8] While the broadband *adoption* rate stands at 28% of U.S. households[9], broadband *availability* is much higher. As of June 30, 2006, the FCC found at least one high-speed subscriber in 99% of all zip codes in the United States. The FCC estimates that "roughly 20 percent of consumers with access to advanced telecommunications capability do subscribe to such services." According to the FCC, possible reasons for the gap between broadband availability and subscribership include the lack of computers in some homes, price of broadband service, lack of content, and the availability of broadband at work.[10]

Broadband in Rural and Underserved Areas[11]

While the number of new broadband subscribers continues to grow, the rate of broadband deployment in urban and high income areas appears to be outpacing deployment in rural and low-income areas. In response to a request by ten Senators, the Departments of Commerce and Agriculture released a report on April 26, 2000, concluding that rural areas lag behind urban areas in access to broadband technology. The report found that less than 5% of towns of 10,000 or less have access to broadband, while broadband over cable has been deployed in more than 65% of all cities with populations over 250,000, and broadband over the telephone network has been deployed in 56% of all cities with populations over 100,000.[12]

Similarly, the February 2002 report from the Department of Commerce, *A Nation Online: How Americans Are Expanding Their Use of the Internet*, found that 12.2% of Internet users in rural areas had high-speed connections, as opposed to 21.2% of Internet users in urban areas. The report's survey also found, not surprisingly, that individuals in high-income households have higher broadband subscribership rates than individuals in lower income households.[13]

December 2005 data from the Pew Internet and American Life Project indicate that while broadband adoption is growing in urban, suburban, and rural areas, broadband users make up larger percentages of urban and suburban users than rural users. Pew found that the percentage of all U.S. adults with broadband at home is 38% for urban areas, 40% for suburban areas, and 24% for rural areas.[14]

Similarly, a May 2006 report released by the Government Accountability Office (GAO) found that 17% of rural households subscribe to broadband, as opposed to 28% of suburban and 29% of urban households.[15]

According to the latest FCC data on the deployment of high-speed Internet connections (released January 2007), high-speed subscribers were reported in 99% of the most densely populated zip codes, as opposed to 89% of zip codes with the lowest population densities. Similarly, for zip codes ranked by median family income, high-speed subscribers were reported present in 99% of the top one-tenth of zip codes, as compared to 91% of the bottom one-tenth of zip codes.[16]

On the other hand, the FCC's *Fourth Report*, while acknowledging that disparities in broadband deployment exist, asserts that the gap between the broadband "haves and have-nots" is narrowing:

> [T]he *Fourth Report* also documents the continuation of a positive trend that first emerged in our last report: namely, the increasing availability of advanced telecommunications capability to certain groups of consumers — those in rural areas, those with low incomes, and those with disabilities — who stand in particular need of advanced services. Consumers in these groups are of special concern to the Commission in that they are most in need of access to advanced telecommunications capability to overcome economic, educational, and other limitations, they are also the most likely to lack access precisely *because* of these limitations. The *Fourth Report* demonstrates that we are making substantial progress in closing the gaps in access that these groups traditionally have experienced.[17]

The September 2004 Department of Commerce report, *A Nation Online: Entering the Broadband Age*, found that a lower percentage of Internet households have broadband in rural areas (24.7%) than in urban areas (40.4%), and that "while broadband usage has grown significantly in all areas since the previous survey, the rural-urban differential continues."[18] The report also found that broadband penetration rates are higher in the West and Northeast than in the South and Midwest.[19] Race and ethnicity were also found to be significant determinants of broadband use, with 25.7% of White Americans living in broadband households, as opposed to 14.2% of Black and 12.6% of Hispanic Americans.[20]

Some policymakers believe that disparities in broadband access across American society could have adverse consequences on those left behind. While a minority of American homes today subscribe to broadband, many believe that advanced Internet applications of the future — voice over the Internet protocol (VoIP) or high quality video, for example — and the resulting ability for businesses and consumers to engage in e-commerce, may increasingly depend on high speed broadband connections to the Internet. Thus, some say, communities and individuals without access to broadband could be at risk to the extent that e-commerce becomes a critical factor in determining future economic development and prosperity. A 2003 study conducted by Criterion Economics found that ubiquitous adoption of current generation broadband technologies would result in a cumulative increase in gross domestic product of $179.7 billion, while sustaining an additional 61,000 jobs per year over the next nineteen years. The study projected that 1.2 million jobs could be created if next generation broadband technology is rapidly and ubiquitously deployed.[21] A February 2006 study done by the Massachusetts Institute of Technology for the Economic Development Administration/ Department of Commerce marked the first attempt to quantitatively measure the impact of broadband on economic growth. The study found that "between 1998 and 2002, communities in which mass-market broadband was available by December 1999 experienced more rapid growth in employment, the number of businesses overall, and businesses in IT-intensive sectors, relative to comparable communities without broadband at that time."[22]

Some also argue that broadband is an important contributor to U.S. future economic strength with respect to the rest of the world. According to the International Telecommunications Union, the U.S. ranks 16th worldwide in broadband penetration (subscriptions per 100 inhabitants as of December 2005).[23] Similarly, data from the Organization for Economic Cooperation and Development (OECD) found the U.S. ranking 12th among OECD nations in broadband access per 100 inhabitants as of June 2006.[24] By contrast, in 2001 an OECD study found the U.S. ranking 4th in broadband subscribership per 100 inhabitants (after Korea, Sweden, and Canada).[25]

Federal Role

The Telecommunications Act of 1996 (P.L. 104-104) addresses the issue of whether the federal government should intervene to prevent a "digital divide" in broadband access. Section 706 requires the FCC to determine whether "advanced telecommunications capability [i.e., broadband or high-speed access] is being deployed to all Americans in a reasonable and timely fashion." If this is not the case, the act directs the FCC to "take immediate action to accelerate deployment of such capability by removing barriers to infrastructure investment and by promoting competition in the telecommunications market."

On January 28, 1999, the FCC adopted its first report (FCC 99-5) pursuant to Section 706. The report concluded that "the consumer broadband market is in the early stages of development, and that, while it is too early to reach definitive conclusions, aggregate data suggests that broadband is being deployed in a reasonable and timely fashion."[26] The FCC announced that it would continue to monitor closely the deployment of broadband capability in annual reports and that, where necessary, it would "not hesitate to reduce barriers to competition and infrastructure investment to ensure that market conditions are conducive to investment, innovation, and meeting the needs of all consumers."

The FCC's second Section 706 report was adopted on August 3, 2000. Based on more extensive data than the first report, the FCC similarly concluded that notwithstanding risks faced by some vulnerable populations, broadband is being deployed in a reasonable and timely fashion overall:

> Recognizing that the development of advanced services infrastructure remains in its early stages, we conclude that, overall, deployment of advanced telecommunications capability is proceeding in a reasonable and timely fashion. Specifically, competition is emerging, rapid build-out of necessary infrastructure continues, and extensive investment is pouring into this segment of the economy.[27]

The FCC's third Section 706 report was adopted on February 6, 2002. Again, the FCC concluded that "the deployment of advanced telecommunications capability to all Americans is reasonable and timely."[28] The FCC added:

> We are encouraged by the expansion of advanced services to many regions of the nation, and growing number of subscribers. We also conclude that investment in infrastructure for most advanced services markets remains strong, even though the pace of investment trends has generally slowed. This may be due in part to the general economic slowdown in the nation. In addition, we find that emerging technologies continue to stimulate competition and create new alternatives and choices for consumers.[29]

On September 9, 2004, the FCC adopted and released its *Fourth Report* pursuant to Section 706. Like the previous three reports, the FCC concludes that "the overall goal of section 706 is being met, and that advanced telecommunications capability is indeed being deployed on a reasonable and timely basis to all Americans."[30] The FCC notes the emergence of new services such as VoIP, and the significant development of new broadband access technologies such as unlicensed wireless (WiFi)and broadband over power lines. The FCC notes the future promise of emerging multiple advanced broadband networks which can complement one another:

> For example, in urban and suburban areas, wireless broadband services may "fill in the gaps" in wireline broadband coverage, while wireless and satellite services may bring high-speed broadband to remote areas where wireline deployment may be costly. Having multiple advanced networks will also promote competition in price, features, and quality-of-service among broadband-access providers.[31]

Two FCC Commissioners (Michael Copps and Jonathan Adelstein) dissented from the *Fourth Report* conclusion that broadband deployment is reasonable and timely. They argued that the relatively poor world ranking of United States broadband penetration indicates that deployment is insufficient, that the FCC's continuing definition of broadband as 200 kilobits per second is outdated and is not comparable to the much higher speeds available to consumers in other countries, and that the use of zip code data (measuring the presence of at least one broadband subscriber within a zip code area) does not sufficiently characterize the availability of broadband across geographic areas.[32]

While the FCC is currently implementing or actively considering some regulatory activities related to broadband,[33] no major regulatory intervention pursuant to Section 706 of the Telecommunications Act of 1996 has been deemed necessary by the FCC at this time.

Meanwhile, the National Telecommunications and Information Administration (NTIA) at the Department of Commerce (DOC) was tasked with developing the Bush Administration's broadband policy.[34] Statements from Administration officials indicated that much of the policy would focus on removing regulatory roadblocks to investment in broadband deployment.[35] On June 13, 2002, in a speech at the 21st Century High Tech Forum, President Bush declared that the nation must be aggressive about the expansion of broadband, and cited ongoing activities at the FCC as important in eliminating hurdles and barriers to get broadband implemented. President Bush made similar remarks citing the economic importance of broadband deployment at the August 13, 2002 economic forum in Waco, Texas.

Subsequently, a more formal Administration broadband policy was unveiled in March and April of 2004. On March 26, 2004, President Bush endorsed the goal of universal broadband access by 2007. Then on April 26, 2004, President Bush announced a broadband initiative which includes promoting legislation which would permanently prohibit all broadband taxes, making spectrum available for wireless broadband and creating technical standards for broadband over power lines, and simplifying rights-of-way processes on federal lands for broadband providers.[36]

The Bush Administration has also emphasized the importance of encouraging demand for broadband services. On September 23, 2002, the DOC's Office of Technology Policy released a report, *Understanding Broadband Demand: A Review of Critical Issues*,[37] which argues that national governments can accelerate broadband demand by taking a number of steps, including protecting intellectual property, supporting business investment, developing e-government applications, promoting efficient radio spectrum management, and others. Similarly, the President's Council of Advisers on Science and Technology (PCAST) was tasked with studying "demand-side" broadband issues and suggesting policies to stimulate broadband deployment and economic recovery. The PCAST report, *Building Out Broadband*, released in December 2002, concludes that while government should not intervene in the telecommunications marketplace, it should apply existing policies and work with the private sector to promote broadband applications and usage. Specific initiatives include increasing e-government broadband applications (including homeland security); promoting telework, distance learning, and telemedicine; pursuing broadband-friendly spectrum policies, and ensuring access to public rights of way for broadband infrastructure.[38] Meanwhile, "high-tech" organizations such as TechNet,[39] the Computer Systems Policy Project (CSPP)[40], and the Semiconductor Industry Association (SIA)[41] have called on the federal government to adopt policies toward a goal of 100 Mbs to 100 million homes by the end of the decade.

Some policymakers in Congress assert that the federal government should play a more active role to avoid a "digital divide" in broadband access, and that legislation is necessary to ensure fair competition and timely broadband deployment. Bills were been introduced into the 109th Congress which seek to provide federal financial assistance for broadband deployment in the form of grants, loans, subsidies, and/or tax credits. Similar bills are expected in the 110th Congress.

State and Local Broadband Activities

In addition to federal support for broadband deployment, there are programs and activities ongoing at the state and local level. Surveys, assessments, and reports from the American Electronics Association,[42] Technet,[43] the Alliance for Public Technology,[44] the California Public Utilities Commission,[45] and the AEI-Brookings Joint Center[46] have explored state and local broadband programs. A related issue is the emergence of municipal broadband networks (primarily wireless and fiber based) and the debate over whether such networks constitute unfair competition with the private sector (for more information on municipal broadband, see CRS Report RS20993, *Wireless Technology and Spectrum Demand: Advanced Wireless Services*, by Linda K. Moore).

FEDERAL TELECOMMUNICATIONS DEVELOPMENT PROGRAMS

Table 1 (at the end of this chapter) shows selected federal domestic assistance programs throughout the federal government that can be associated with telecommunications development. Many (if not most) of these programs can be related, if not necessarily to the deployment of broadband technologies in particular, then to telecommunications and the "digital divide" issue generally.

The Universal Service Concept and the FCC[47]

Since its creation in 1934 the Federal Communications Commission (FCC) has been tasked with "... mak[ing] available, so far as possible, to all the people of the United States, ...

a rapid, efficient, Nation-wide, and world-wide wire and radio communications service with adequate facilities at reasonable charges...."[48] This mandate led to the development of what has come to be known as the universal service concept.

The universal service concept, as originally designed, called for the establishment of policies to ensure that telecommunications services are available to all Americans, including those in rural, insular and high cost areas, by ensuring that rates remain affordable. Over the years this concept fostered the development of various FCC policies and programs to meet this goal. The FCC offers universal service support through a number of direct mechanisms that target both providers of and subscribers to telecommunications services.[49]

The development of the federal universal service high cost fund is an example of provider-targeted support. Under the high cost fund, eligible telecommunications carriers, usually those serving rural, insular and high cost areas, are able to obtain funds to help offset the higher than average costs of providing telephone service.[50] This mechanism has been particularly important to rural America where the lack of subscriber density leads to significant costs. FCC universal service policies have also been expanded to target individual users. Such federal programs include two income-based programs, Link Up and Lifeline, established in the mid-1980s to assist economically needy individuals. The Link Up program assists low-income subscribers pay the costs associated with the initiation of telephone service and the Lifeline program assists low-income subscribers pay the recurring monthly service charges. Funding to assist carriers providing service to individuals with speech and/or hearing disabilities is also provided through the Telecommunications Relay Service Fund. Effective January 1, 1998, schools, libraries, and rural health care providers also qualified for universal service support.

Universal Service and the Telecommunications Act of 1996

Passage of the Telecommunications Act of 1996 (P.L.104-104) codified the long-standing commitment by U.S. policymakers to ensure universal service in the provision of telecommunications services.

The Schools and Libraries, and Rural Health Care Programs

Congress, through the 1996 Act, not only codified, but also expanded the concept of universal service to include, among other principles, that elementary and secondary schools and classrooms, libraries, and rural health care providers have access to telecommunications services for specific purposes at discounted rates. (See Sections 254(b)(6) and 254(h)of the 1996 Telecommunications Act, 47 USC 254.)

1. The Schools and Libraries Program

Under universal service provisions contained in the 1996 Act, elementary and secondary schools and classrooms and libraries are designated as beneficiaries of universal service discounts. Universal service principles detailed in Section 254(b)(6) state that "Elementary and secondary schools and classrooms ... and libraries should have access to advanced telecommunications services..." The act further requires in Section 254(h)(1)(B) that services within the definition of universal service be provided to elementary and secondary schools and libraries for education purposes at discounts, that is at "rates less than the amounts charged for similar services to other parties."

The FCC established the Schools and Libraries Division within the Universal Service Administrative Company (USAC) to administer the schools and libraries or "E (education)-rate" program to comply with these provisions. Under this program, eligible schools and libraries receive discounts ranging from 20 to 90 percent for telecommunications services depending on the poverty level of the school's (or school district's) population and its location in a high cost telecommunications area. Three categories of services are eligible for discounts: internal connections (e.g. wiring, routers and servers); Internet access; and telecommunications and dedicated services, with the third category receiving funding priority. According to data released by program administrators, $17 billion in funding has been committed over the first eight years of the program with funding released to all states, the District of Columbia and all territories. Funding commitments for funding Year 2006 (July 1, 2006 - June 30, 2007), the ninth and current year of the program, totaled $1.6 billion as of December 5, 2006.[51]

2. The Rural Health Care Program

Section 254(h) of the 1996 Act requires that public and non-profit rural health care providers have access to telecommunications services necessary for the provision of health care services at rates comparable to those paid for similar services in urban areas. Subsection 254(h)(1) further specifies that "to the extent technically feasible and economically reasonable" health care providers should have access to advanced telecommunications and information services. The FCC established the Rural Health Care Division (RHCD) within the USAC to administer the universal support program to comply with these provisions. Under FCC established rules only public or nonprofit health care providers are eligible to receive funding. Eligible health care providers, with the exception of those requesting only access to the Internet, must also be located in a rural area.[52] The funding ceiling, or cap, for this support was established at $400 million annually. The funding level for Year One of the program (January 1998 - June 30, 1999) was set at $100 million. Due to less than anticipated demand, the FCC established a $12 million funding level for the second year (July 1, 1999 to June 30, 2000) of the program but has since returned to a $400 million yearly cap. As of December 8, 2006, covering the first nine years of the program, a total of $174.8 million has been committed to 3,440 rural health care providers. The primary use of the funding is to provide reduced rates for telecommunications and information services necessary for the provision of health care.[53]

The Telecommunications Development Fund

Section 714 of the 1996 Act created the Telecommunications Development Fund (TDF). The TDF is a private, non-governmental, venture capital corporation overseen by a seven-member board of directors and fund management. The purpose of the TDF is threefold: to promote access to capital for small businesses in order to enhance competition in the telecommunications industry; to stimulate new technology development and promote employment and training; and to support universal service and enhance the delivery of telecommunications services to rural and underserved areas. The TDF is authorized to provide financing to eligible small businesses in the telecommunications industry through loans and investment capital. At this time the TDF is focusing on providing financing in the form of equity investments ranging from $375,000 to $1 million per investment.[54] Initial funding for the program is derived from the interest earned from the upfront payments bidders submit to participate in FCC auctions. The availability of funds for future investments is dependent on earning a successful return on the Fund's portfolio. As of September 2005, the TDF had $50 million under management of which $16.8million is committed to seventeen portfolio companies. [55]

Universal Service and Broadband

One of the policy debates surrounding universal service is whether access to advanced telecommunications services (i.e. broadband) should be incorporated into universal service objectives. The term universal service, when applied to telecommunications, refers to the ability to make available a basket of telecommunications services to the public, across the nation, at a reasonable price. As directed in the 1996 Telecommunications Act [Section 254(c)] a federal-state Joint Board was tasked with defining the services which should be included in the basket of services to be eligible for federal universal service support; in effect using and defining the term "universal service" for the first time. The Joint Board's recommendation, which was subsequently adopted by the FCC in May 1997, included the following in its universal services package: voice grade access to and some usage of the public switched network; single line service; dual tone signaling; access to directory assistance; emergency service such as 911; operator services; access and interexchange (long distance) service.

Some policy makers expressed concern that the FCC-adopted definition is too limited and does not take into consideration the importance and growing acceptance of advanced services such as broadband and Internet access. They point to a number of provisions contained in the Universal Service section of the 1996 Act to support their claim. Universal service principles contained in Section 254(b)(2) state that "Access to advanced telecommunications services should be provided to all regions of the Nation." The subsequent principle (b)(3) calls for consumers in all regions of the Nation including "low-income" and those in "rural, insular, and high cost areas" to have access to telecommunications and information services including "advanced services" at a comparable level and a comparable rate charged for similar services in urban areas. Such provisions, they state, dictate that the FCC expand its universal service definition.

Others caution that a more modest approach is appropriate given the "universal mandate" associated with this definition and the uncertainty and costs associated with mandating nationwide deployment of such advanced services as a universal service policy goal. Furthermore they state the 1996 Act does take into consideration the changing nature of the telecommunications sector and allows for the universal service definition to be modified if future conditions warrant. Section 254(c)of the act states that "universal service is an evolving level of telecommunications services" and the FCC is tasked with "periodically" reevaluating this definition "taking into account advances in telecommunications and information technologies and services." Furthermore, the Joint Board is given specific authority to recommend "from time to time" to the FCC modification in the definition of the services to be included for federal universal service support. The Joint Board, in July 2002, concluded such an inquiry and recommended that at this time no changes be made in the current list of services eligible for universal service support. The FCC, in a July 10, 2003 order (FCC 03-170) adopted the Joint Board's recommendation thereby leaving unchanged the list of services supported by Federal universal service.

Rural Utilities Service

The Rural Electrification Administration (REA), subsequently renamed the Rural Utilities Service (RUS), was established by the Roosevelt Administration in 1935. Initially, it was established to provide credit assistance for the development of rural electric systems. In 1949, the mission of REA was expanded to include rural telephone providers. Congress further amended the Rural Electrification Act in 1971 to establish within REA a Rural Telephone Account and the Rural Telephone Bank (RTB). Rural Telephone Loans and Loan Guarantees provide long-term direct and guaranteed loans for telephone lines, facilities, or systems to furnish and improve telecommunications service in rural areas. The RTB – liquidated in FY2006 – was a public-private partnership intended to provide additional sources of capital that would supplement loans made directly by RUS. Another program, the Distance Learning and Telemedicine Program, specifically addresses health care and education needs of rural America.

RUS implements two programs specifically targeted at providing assistance for broadband deployment in rural areas: the Rural Broadband Access Loan and Loan Guarantee Program and Community Connect Broadband Grants. The current authorization for the Rural Broadband Access Loan and Loan Guarantee Program expires on September 30, 2007. It is expected that the 110th Congress will consider reauthorization of the program as part of the 2007 farm bill. For further information on rural broadband and the RUS broadband programs, see CRS Report RL33816, *Broadband Loan and Grant Programs in the USDA's Rural Utilities Service, by* Lennard G. Kruger.

LEGISLATION IN THE 109TH CONGRESS

In the 109th Congress, legislation was introduced to provide financial assistance to encourage broadband deployment (including loans, grants, and tax incentives), and to allocate additional spectrum for use by wireless broadband applications. Of particular note is enactment of the Deficit Reduction Act of 2005 (P.L. 109-171), which set a hard deadline for the digital television transition, thereby reclaiming analog television spectrum to be auctioned for commercial applications such as wireless broadband. A complete listing of bills is provided below.

H.R. 3 (Young, Don)/P.L. 109-59

Transportation Equity Act: A Legacy for Users. Directs the Secretary of Transportation to conduct a study on the feasibility of installing fiber optic cabling and wireless communications infrastructure along rural interstate highway corridors; such study will identify rural broadband access points. Introduced February 9, 2005; referred to Committee on Transportation and Infrastructure. Passed House on March 10, 2005. Passed Senate on May 17, 2005. Signed into law by President on August 10, 2005.

H.R. 144 (McHugh)

Rural America Digital Accessibility Act. Provides for grants, loans, research, and tax credits to promote broadband deployment in underserved rural areas. Introduced January 4,

2005; referred to Committee on Energy and Commerce and the Committee on Ways and Means.

H.R. 146 (McHugh)

Establishes a grant program to support broadband-based economic development efforts. Introduced January 4, 2005; referred to Committee on Transportation and Infrastructure and to Committee on Financial Services.

H.R. 1479 (Udall)

Rural Access to Broadband Service Act. Establishes a Rural Broadband Office within the Department of Commerce which would coordinate federal government resources with respect to expansion of broadband services in rural areas. Directs the National Science Foundation to conduct research in enhancing rural broadband. Expresses the Sense of Congress that the broadband loan program in the Rural Utilities Service should be fully funded. Provides for the expensing of broadband Internet access expenditures for rural communities. Introduced April 5, 2005; referred to Committees on Science and on Energy and Commerce.

H.R. 3517 (Andrews)

Greater Access to E-Governance Act (GATE Act). Establishes a grant program in the Department of Commerce to provide funds to State and local governments to enable them to deploy broadband computer networks for the conduct of electronic governance transactions by citizens in local schools and libraries. Introduced July 28, 2005; referred to Committee on Energy and Commerce.

H.R. 3958 (Melancon)

Louisiana Katrina Reconstruction Act. Provides grants for construction of broadband infrastructure necessary for technology and economic development in areas affected by Hurricane Katrina. Introduced September 29, 2005; referred to multiple committees.

H.R. 5072 (Terry)

Universal Reform Act of 2006. Targets universal service support specifically to eligible telecommunications carriers in high-cost geographic areas to ensure that communications services and high-speed broadband services are made available throughout all of the States of the United States in a fair and equitable manner. Introduced March 30, 2006; referred to Committee on Energy and Commerce.

H.R. 5252 (Barton)

Communications, Opportunity, Promotion, and Enhancement Act of 2006. Establishes Broadband for Unserved Areas Program funded by universal service funding not to exceed $500 million per year (Title II). Prevents states from prohibiting municipal broadband (Title V). Directs FCC to make available eligible broadcast television frequencies ("white space") for unlicensed use (which could include wireless broadband applications) in a manner that protects broadcasters from interference (Title VI). Directs the FCC to periodically revise its definition of broadband above 200 kbps (Title IX). Requires the FCC to collect more detailed broadband deployment data (Title X). Introduced May 1, 2006. Passed House June 8, 2006.

Reported by Senate Committee on Commerce, Science and Transportation with an amendment in the nature of a substitute, September 29, 2006 (S.Rept. 109-355).

H.R. 5970 (Thomas, William)

Estate Tax and Extension of Tax Relief Act of 2006. Provides a tax credit to holders of rural renaissance bonds funding qualified projects including expanding broadband technology in rural areas. Passed House July 29, 2006.

S. 14 (Stabenow)

Fair Wage, Competition, and Investment Act of 2005. Allows the expensing of broadband Internet access expenditures. Introduced January 24, 2005; referred to Committee on Finance.

S. 497 (Salazar)

Broadband Rural Revitalization Act of 2005. Establishes a Rural Broadband Office within the Department of Commerce which would coordinate federal government resources with respect to expansion of broadband services in rural areas. Expresses the Sense of Congress that the broadband loan program in the Rural Utilities Service should be fully funded. Provides for the expensing of broadband Internet access expenditures for rural communities. Introduced March 2, 2005; referred to Committee on Finance.

S. 502 (Coleman)

Rural Renaissance Act. Creates a Rural Renaissance Corporation which would fund qualified projects including projects to expand broadband technology in rural areas. Introduced March 3, 2005; referred to Committee on Finance.

S. 1147 (Rockefeller)

Amends the Internal Revenue Code of 1986 to provide for the expensing of broadband Internet access expenditures. Introduced May 26, 2005; referred to Committee on Finance.

S. 1321 (Santorum)

Telephone Excise Tax Repeal Act of 2005. Amends the Internal Revenue Code of 1986 to provide for the expensing of broadband Internet access expenditures. Introduced June 28, 2005; reported by Committee on Finance, September 15, 2006 (S.Rept. 109-336).

S. 1583 (Smith)

Universal Service for the 21st Century Act. Amends the Communications Act of 1934 to expand the contribution base for universal service and to establish a separate account — not to exceed $500 million per year — within the universal service fund to support the deployment of broadband service in unserved areas of the United States. Introduced July 29, 2005; referred to Committee on Commerce, Science and Transportation.

S. 1932 (Gregg)

Deficit Reduction Act of 2005. Section 1401 cancels unobligated funds remaining as of October 1, 2006 for the USDA Rural Utilities Service Rural Broadband Access Loan and Loan Guarantee Program. Title III sets a hard deadline for the digital television transition, thereby reclaiming analog television spectrum to be auctioned for commercial applications such as wireless broadband. Passed Senate, November 3, 2005. House agreed to conference report (H.Rept. 109-362), December 19, 2005. Senate agreed to conference report with amendments, December 21, 2005. House agreed to amended conference report, February 1, 2006. **P.L. 109-171** signed by President, February 8, 2006.

S. 1765 (Landrieu)

Louisiana Katrina Reconstruction Act. Provides grants for construction of broadband infrastructure necessary for technology and economic development in areas affected by Hurricane Katrina. Introduced September 22, 2005; referred to Committee on Finance.

S. 1766 (Vitter)

Louisiana Katrina Reconstruction Act. Provides grants for construction of broadband infrastructure necessary for technology and economic development in areas affected by Hurricane Katrina. Introduced September 22, 2005; referred to Committee on Finance.

S. 2020 (Grassley)

Tax Relief Act of 2005. Provides a tax credit to holders of rural renaissance bonds funding qualified projects including expanding broadband technology in rural areas. Passed by Senate as H.R. 4297, February 2, 2006. Provision not retained in Conference Report.

S. 2256 (Burns)

Internet and Universal Service Act of 2006. Amends the Communications Act of 1934 to ensure the availability to all Americans of high-quality, advanced telecommunications and broadband services, technologies, and networks at just, reasonable, and affordable rates, and to establish a permanent mechanism to guarantee specific, sufficient, and predictable support for the preservation and advancement of universal service. Introduced February 8, 2006; referred to Committee on Commerce, Science, and Transportation.

S. 2357 (Kennedy)

Right TRACK Act. Directs the President's Council of Advisors on Science and Technology to establish a national broadband policy for improving and expanding broadband access in the United States by 2010. Introduced March 2, 2006; referred to Committee on Finance.

S. 2686 (Stevens)

Communications, Consumer's Choice, and Broadband Deployment Act of 2006. A bill to amend the Communications Act of 1934 and for other purposes. Includes provision on universal service reform. Introduced May 1, 2006; referred to Committee on Commerce, Science, and Transportation.

S. 3820 (Durbin)

Broadband for Rural America Act of 2006. Establishes a Broadband Access Trust Fund and Office of Broadband Access within the FCC to provide grants to study the lack of affordable broadband in unserved areas. Also reforms FCC's broadband data reporting and USDA's broadband loan and grant programs, provides for spectrum auction for wireless rural broadband, and establishes a public-private Rural Broadband Access Task Force. Introduced August 3, 2006; referred to Committee on Commerce, Science, and Transportation.

S. 3829 (Stabenow)

Tax Relief and Minimum Wage Act of 2006. Provides a tax credit to holders of rural renaissance bonds funding qualified projects including expanding broadband technology in rural areas. Introduced August 3, 2006; referred to Committee on Finance.

S. 3936 (Frist)

National Competitiveness Investment Act. Authorizes the National Science Foundation to provide grants for basic research in advanced information and communications technologies. Areas of research include affordable broadband access, including wireless technologies. Introduced September 26, 2006; placed on Senate Legislative Calendar.

S. 3999 (Clinton)

Rural Broadband Initiative Act of 2006. Establishes an Office of Rural Broadband Initiatives within the Department of Agriculture which will administer all rural broadband grant and loan programs previously administered by the Rural Utilities Service. Also establishes a National Rural Broadband Innovation Fund which would fund experimental and pilot rural broadband projects and applications. Introduced September 29, 2006; placed on Senate Legislative Calendar.

LEGISLATION IN THE 110TH CONGRESS

Many of the legislative proposals related to providing financial assistance for broadband deployment are likely to be reintroduced into the 110th Congress. Of particular note is the possible reauthorization of the Rural Utilities Service (RUS) broadband program, which is expected to be considered as part of the 2007 farm bill. Legislation to reform universal service – which could have a significant impact on the amount of financial assistance available for broadband deployment in rural and underserved areas – has been introduced into the 110th Congress (H.R. 42, S. 101).

H.R. 42 (Velazquez)

Serving Everyone with Reliable, Vital Internet, Communications and Education Act of 2007. Directs the FCC to expand assistance provided by the Lifeline Assistance Program and the Link Up Program to include broadband service. Introduced January 4, 2007; referred to Committee on Energy and Commerce.

S. 101 (Stevens)

Universal Service for Americans Act ("USA Act"). Directs the FCC to establish Broadband for Unserved Area Areas Program to be funded by the Universal Service Fund. Requires communications carriers to submit detailed broadband deployment data to the FCC. Introduced January 4, 2007; referred to Committee on Commerce, Science, and Transportation.

POLICY ISSUES

Legislation in the 110th Congress will likely seek to provide federal financial assistance for broadband deployment in rural and underserved areas. In assessing this legislation, several policy issues arise.

Is Broadband Deployment Data Adequate?

Obtaining an accurate snapshot of the status of broadband deployment is problematic. Anecdotes abound of rural and low-income areas which do not have adequate Internet access, as well as those which are receiving access to high-speed, state-of-the-art connections. Rapidly evolving technologies, the constant flux of the telecommunications industry, the uncertainty of consumer wants and needs, and the sheer diversity and size of the nation's economy and geography make the status of broadband deployment very difficult to characterize. The FCC periodically collects broadband deployment data from the private sector via "FCC Form 477" — a standardized information gathering survey. Statistics derived from the Form 477 survey are published every six months. Additionally, data from Form 477 are used as the basis of the FCC's (to date) four broadband deployment reports. The FCC is working to refine the data used in future Reports in order to provide an increasingly accurate portrayal. In its March 17, 2004 Notice of Inquiry for the *Fourth Report*, the FCC sought comments on specific proposals to improve the FCC Form 477 data gathering program.[56] On November 9, 2004, the FCC voted to expand its data collection program by requiring reports from all facilities based carriers regardless of size in order to better track rural and underserved markets, by requiring broadband providers to provide more information on the speed and nature of their service, and by establishing broadband-over-power line as a separate category in order to track its development and deployment. The FCC Form 477 data gathering program is extended for five years beyond its March 2005 expiration date.[57]

The Government Accountability Office (GAO) has cited concerns about the FCC's zip-code level data. Of particular concern is that the FCC will report broadband service in a zip code even if a company reports service to only one subscriber, which in turn can lead to some observers overstating of broadband deployment. According to GAO, "the data may not provide a highly accurate depiction of local deployment of broadband infrastructures for residential service, especially in rural areas." The FCC has acknowledged the limitations in its zip code level data.[58]

In the 109th Congress, H.R. 5252, the communications reform bill as marked up by the Senate Committee on Commerce, Science and Transportation on June 28, 2006, contained a provision which sought to revise the FCC's broadband data collection program. Section 1011, "Broadband Reporting Requirements," would have required the FCC to collect data by "zip code plus four" service areas. Specifically, the FCC would require data for each service area on the percentage of households offered broadband service, the percentage of households actually subscribing to broadband service, the average price per megabyte of download and upload speeds, actual average throughput speeds, and the ratio of the number of users sharing the same line. Exemptions would be allowed if the FCC determined that a broadband provider's compliance with the reporting requirements was cost prohibitive. Additionally, the

Table 1. Selected Federal Domestic Assistance Programs Related to Telecommunications Development

Program	Agency	Description	FY2006 obligations)	Web Links for More Information [http://12.46.245.173/cfda/cfda.html]: Go to "All Programs Listed Numerically" and search by program
Public Telecommunications Facilities — Planning and Construction	National Telecommunications and Information Administration, Dept. of Commerce	Assists in planning, acquisition, installation and modernization of public telecommunications facilities	$19.7 million	http://www.ntia.doc.gov/otiahome/ptfp/index.html]
Grants for Public Works and Economic Development Facilities	Economic Development Administration, Dept. of Commerce	Provides grants to economically distressed areas for construction of public facilities and infrastructure, including broadband deployment and other types of telecommunications enabling projects	$158.1 million	[http://www.eda.gov/]
Rural Telephone Loans and Loan Guarantees	Rural Utilities Service, U.S. Dept. of Agriculture	Provides long-term direct and guaranteed loans to qualified organizations for the purpose of financing the improvement, expansion, construction, acquisition, and operation of telephone lines, facilities, or systems to furnish and improve telecommunications service in rural areas	$145 million (hardship loans); $420 million (cost of money loans); $175 million (FFB Treasury loans)	[http://www.usda.gov/rus/telecom/index.htm]

Table 1. (Continued).

Program	Agency	Description	FY2006 obligations)	Web Links for More Information [http://12.46.245.173/cfda/cfda.html]: Go to "All Programs Listed Numerically" and search by program
Distance Learning and Telemedicine Loans and Grants	Rural Utilities Service, U.S. Dept. of Agriculture	Provides seed money for loans and grants to rural community facilities (e.g., schools, libraries, hospitals) for advanced telecommunications systems that can provide health care and educational benefits to rural areas	$54.4 million (grants) $25 million (loans)	[http://www.usda.gov/rus/telecom/dlt/dlt.htm]
Rural Broadband Access Loan and Loan Guarantee Program	Rural Utilities Service, U.S. Dept. of Agriculture	Provides loan and loan guarantees for facilities and equipment providing broadband service in rural communities	$2032 million (cost of money loan) $46 million (4% loan) $79 million (loan guarantee)	[http://www.usda.gov/rus/telecom/broadband.htm]
Community Connect Broadband Grants	Rural Utilities Service, U.S. Dept. of Agriculture	Provides grants to applicants proposing to provide broadband service on a "communityoriented connectivity" basis to rural communities of under 20,000 inhabitants.	$9 million	[http://www.usda.gov/rus/telecom/index.htm]
Education Technology State Grants	Office of Elementary and Secondary Education, Dept. of Education	Grants to State Education Agencies for development of information technology to improve teaching and learning in schools	$272 million	[http://www.ed.gov/Technology/TLCF/index.html]

Table 1. (Continued).

Program	Agency	Description	FY2006 obligations)	Web Links for More Information [http://12.46.245.173/cfda/cfda.html]: Go to "All Programs Listed Numerically" and search by program
Star Schools	Office of Assistant Secretary for Educational Research and Improvement, Dept. of Education	Grants to telecommunication partnerships for telecommunications facilities and equipment, educational and instructional programming	$14.8 million	[http://www.ed.gov/programs/starschools/index.html]
Ready to Teach	Office of Assistant Secretary for Educational Research and Improvement, Dept. of Education	Grants to carry out a national telecommunication-based program to improve the teaching in core curriculum areas.	$10.9 million	[http://www.ed.gov/programs/readyteach/index.html]
Special Education — Technology and Media Services for Individuals with Disabilities	Office of Special Education and Rehabilitative Services, Dept. of Education	Supports development and application of technology and education media activities for disabled children and adults	$38.4 million	[http://www.ed.gov/about/offices/list/osers/index.html?src=mr/]
Telehealth Network Grants	Health Resources and Services Administration, Department of Health and Human Services	Grants to develop sustainable telehealth programs and networks in rural and frontier areas, and in medically unserved areas and populations.	$3.4 million	[http://www.hrsa.gov/telehealth/]
Medical Library Assistance	National Library of Medicine, National Institutes of Health, Department of Health and Human Services	Provides funds to train professional personnel; strengthen library and information services; facilitate access to and delivery of health science information; plan and develop advanced information networks; support certain kinds of biomedical publications; and conduct research in medical informatics and related sciences	$65.2 million	[http://www.nlm.nih.gov/ep/extramural.html]

Table 1. (Continued).

Program	Agency	Description	FY2006 obligations)	Web Links for More Information [http://12.46.245.173/cfda/cfda.html]: Go to "All Programs Listed Numerically" and search by program
State Library Program	Office of Library Services, Institute of Museum and Library Services, National Foundation on the Arts and the Humanities	Grants to state library administrative agencies for promotion of library services that provide all users access to information through State, regional, and international electronic networks	$163.7 million	[http://www.imls.gov/grants/library/lib_gsla.asp#po]
Native American and Native Hawaiian Library Services	Office of Library Services, Institute of Museum and Library Services, National Foundation on the Arts and the Humanities	Supports library services including electronically linking libraries to networks	$3.6 million	[http://www.imls.gov/grants/library/lib_nat.asp]
Appalachian Area Development	Appalachian Regional Commission	Provides project grants for Appalachian com-munities to support the physical infrastructure necessary for economic development and improved quality of life.	$62 million	[http://www.arc.gov/index.do?nodeId=21]
Denali Commission Program	Denali Commission	Provides grants through a federal and state partnership designed to provide critical infrastructure and utilities throughout Alaska, particularly in distressed communities	$139 million	[http://www.denali.gov/]

Prepared by CRS based on information from the Catalog of Federal Domestic Assistance, updated December 2006.

FCC would be directed to provide to Congress an annual report on the demographics of areas not served by any broadband provider.

Is Federal Assistance for Broadband Deployment Premature or Inappropriate?

Related to the data issue is the argument that government intervention in the broadband marketplace would be premature or inappropriate. Some argue that financial assistance for broadband deployment could distort private sector investment decisions in a dynamic and rapidly evolving marketplace, and question whether federal tax dollars should support a technology that has not yet matured, and whose societal benefits have not yet been demonstrated.[59]

On the other hand, proponents of financial assistance counter that the available data show, in general, that the private sector will invest in areas where it expects the greatest return — areas of high population density and income. Without some governmental assistance in underserved areas, they argue, it is reasonable to conclude that broadband deployment will lag behind in many rural and low income areas.[60]

Which Approach Is Best?

If one assumes that governmental action is appropriate to spur broadband deployment in underserved areas, which specific approaches, either separately or in combination, would likely be most effective? Targeted grants and loans from several existing federal programs have been proposed, as well as tax credits for companies deploying broadband systems in rural and low-income areas. How might the impact of federal assistance compare with the effects of regulatory or deregulatory actions?[61] And finally, how might any federal assistance programs best compliment existing "digital divide" initiatives by the states, localities, and private sector?

References

[1] The term "digital divide" can also refer to international disparities in access to information technology. This chapter focuses on domestic issues only.

[2] See U.S. Department of Commerce, *Falling Through the Net: Toward Digital Inclusion*, released October 2000.

[3] Not all observers agree that a "digital divide" exists. See, for example: Thierer, Adam D., *Divided Over the Digital Divide*, Heritage Foundation, March 1, 2000. [http://www.heritage.org/Press/ Commentary/ED030100.cfm]

[4] Department of Commerce, *A Nation Online: How Americans Are Expanding Their Use of the Internet*, February 2002. Based on a September 2001 Census Bureau survey of 57,000 households. See [http://www.ntia.doc.gov/ntiahome/dn/index.html]

[5] A Nation Online, pp. 10-11.

[6] For further information on different types of broadband technologies, including their respective strengths and limitations, see CRS Report RL33542, *Broadband Internet Access: Background and Issues*, by Angele A. Gilroy and Lennard G. Kruger.

[7] U.S. Department of Commerce, Economics and Statistics Administration, National Telecommunications and Information Administration, *A Nation Online: Entering the Broadband Age*, September 2004, p. 1.

[8] FCC, *High-Speed Services for Internet Access: Status as of June 30, 2006,* January 2007. Available at [http://hraunfoss.fcc.gov/edocs_public/attachmatch/DOC-270128A1.pdf]

[9] U.S. Government Accountability Office, Broadband Deployment is Extensive throughout the United States, but It Is Difficult to Assess the Extent of Deployment Gaps in Rural Areas, GAO-06-426, May 2006, p. 3.

[10] Federal Communications Commission, *Fourth Report to Congress*, "Availability of Advanced Telecommunications Capability in the United States," GN Docket No. 04-54, FCC 04-208, September 9, 2004, p. 38. Available at [http://hraunfoss.fcc.gov/edocs_public/attachmatch/FCC-04-208A1.pdf]

[11] For more information on rural broadband and broadband programs at the Rural Utilities Service, see CRS Report RL33816, *Broadband Loan and Grant Programs in the USDA's Rural Utilities Service, by* Lennard G. Kruger.

[12] See U.S. Depts. of Commerce and Agriculture, *Advanced Telecommunications in Rural America: The Challenge of Bringing Broadband Service to All Americans*, April 2000, 80 pages. Available at [http://www.ntia.doc.gov/reports/ruralbb42600.pdf]

[13] A Nation Online, pp. 40-41.

[14] Horrigan, John B., Pew Internet and American Life Project, *Rural Broadband Internet Use,* February 2006, Available at [http://www.pewinternet.org/ pdfs/PIP_Rural_Broadband.pdf]

[15] U.S. Government Accountability Office, Broadband Deployment is Extensive throughout the United States, but It Is Difficult to Assess the Extent of Deployment Gaps in Rural Areas, GAO-06-426, May 2006, p. 12.

[16] FCC, *High-Speed Services for Internet Access: Status as of June 30, 2006,* January 2007, p. 4. Available at [http://hraunfoss.fcc.gov/ edocs_public/attachmatch/DOC-270128A1.pdf]

[17] Fourth Report, p. 8-9.

[18] A Nation Online: Entering the Broadband Age, pp. 12-13.

[19] *Ibid.*, p. 12.

[20] *Ibid.*, p. A-1.

[21] Crandall, Robert W. et al, *The Effect of Ubiquitous Broadband Adoption on Investment, Jobs, and the U.S. Economy*, Conducted by Criterion Economics, L.L.C. for the New Millennium Research Council, September 2003. Available at [http://www. Newmillenniumresearch.org/ archive/bbstudyreport_091703.pdf]

[22] Gillett, Sharon E., Massachusetts Institute of Technology, "Measuring Broadband's Economic Impact," report prepared for the Economic Development Administration, U.S. Department of Commerce, February 28, 2006 p. 4.

[23] International Telecommunications Union, *Economies by broadband penetration, 2005.* Available at [http://www.itu.int/ITU-D/ict/statistics/at_glance/top20_broad_2005.html].

[24] OECD, *OECD Broadband Statistics, June 2006*. Available at [http://www.oecd.org/document/9/0,2340,en_2649_34223_37529673_1_1_1_1,00.html]
[25] OECD, Directorate for Science, Technology and Industry, *The Development of Broadband Access in OECD Countries*, October 29, 2001, 63 pages. For a comparison of government broadband policies, also see OECD, Directorate for Science, Technology and Industry, *Broadband Infrastructure Deployment: The Role of Government Assistance*, May 22, 2002, 42 pages.
[26] FCC News Release, "FCC Issues Report on the Deployment of Advanced Telecommunications Capability to All Americans," January 28, 1999. [http://www.fcc.gov/Bureaus/Common_Carrier/News_ Releases/1999/nrcc9004.html]
[27] Deployment of Advanced Telecommunications Capability: Second Report, p. 6.
[28] Third Report, p. 5.
[29] *Ibid.*, p. 5-6.
[30] Fourth Report, p. 8.
[31] *Ibid.*, p. 9.
[32] *Ibid.*, p. 5, 7.
[33] See Appendix C of the *Fourth Report*, "List of Broadband-Related Proceedings at the Commission," pp. 54-56.
[34] See speech by Nancy Victory, Assistant Secretary for Communications and Information, before the National Summit on Broadband Deployment, October 25, 2001, [http://www.ntia.doc.gov/ntiahome/ speeches/2001/broadband_102501.htm]
[35] Address by Nancy Victory, NTIA Administrator, before the Alliance for Public Technology Broadband Symposium, February 8, 2002, [http://www.ntia.doc.gov/ntiahome/speeches/2002/apt_020802.htm]
[36] See White House, *A New Generation of American Innovation*, April 2004. Available at [http://www.whitehouse.gov/infocus/ technology/economic_policy200404/innovation.pdf]
[37] Available at [http://www.technology.gov/reports/TechPolicy/Broadband _020921.pdf]
[38] President's Council of Advisors on Science and Technology, Office of Science and Technology Policy, *Building Out Broadband*, December 2002, 14 p. Available at [http://www.ostp.gov/PCAST/FINAL%20 Broadband%20Report%20With%20Letters.pdf]
[39] TechNet represents over 300 senior executives from companies in the fields of information technology, biotechnology, venture capital, investment banking, and law. TechNet's policy document, "A National Imperative: Universal Availability of Broadband by 2010," is available at [http://www.technet.org/news/release/?postId=6265 and pageTitle=TechNet+CEOs+Call+for+National+Broadband+Policy]
[40] CSPP is composed of nine CEOs from computer hardware and information technology companies. See "A Vision for 21st Century Wired and Wireless Broadband: Building the Foundation of the Networked World," [http://www.cspp.org/documents /networkedworld.pdf]
[41] See Semiconductor Industry Association, "Removing Barriers to Broadband Deployment," [http://www.sia-online.org/downloads/ Broadband_Combined.pdf]
[42] American Electronics Association, *Broadband in the States 2003: A State-by-State Overview of Broadband Deployment*, May 22, 2003. [http://www.aeanet.org/publications/idet_broadbandstates03.asp]

[43] TechNet, *The State Broadband Index: An Assessment of State Policies Impacting Broadband Deployment and Demand*, July 17, 2003, 48 p. Available at [http://www.michigan.gov/documents/State_Broadband_Index_71282_7.pdf]

[44] Alliance for Public Technology, *A Nation of Laboratories: Broadband Policy Experiments in the States*, March 5, 2004, 48 p. Available at [http://www.apt.org /publications/reports-studies/broadbandreport _final.pdf]

[45] California Public Utilities Commission, *Broadband Deployment in California*, May 5, 2005, 83 p. Available at [http://www.cpuc. ca.gov/static/telco/reports/ broad band report.htm]

[46] Wallsten, Scott, AEI-Brookings Joint Center for Regulatory Studies, *Broadband Penetration: An Empirical Analysis of State and Federal Policies*, Working Paper 05-12, June 2005, 29 p. Available at[http://aei-brookings.org/admin/authorpdfs/ page. php?id=1161]

[47] The section on universal service was prepared by Angele Gilroy, Specialist in Telecommunications, Resources, Science and Industry Division.

[48] Communications Act of 1934, As Amended, Title I sec.1[47 U.S.C. 151].

[49] Many states participate in or have programs that mirror FCC universal service mechanisms to help promote universal service goals within their states.

[50] Additional FCC policies such as rate averaging and pooling have also been implemented to assist high cost carriers.

[51] For additional information on this program, including funding commitments, see the E-rate website: [http://www.universal service.org/sl/]

[52] Any health care provider that does not have toll-free access to the Internet can receive the lesser of $180 in toll charges per month or the toll charges incurred for 30 hours of access to the Internet per month. To obtain this support the health care provider does not have to be located in a rural area, but must show that it lacks toll-free Internet access and that it is an eligible health care provider.

[53] For additional information on this program, including funding commitments, see the RHCD website: [http://www.universal service.org/rhc/]

[54] The TDF also provides management and technical assistance to the companies in which it invests.

[55] For additional information on this program see the TDF website at [http://www. tdfund.com]

[56] Federal Communications Commission, *Notice of Inquiry*, "Concerning the Deployment of Advanced Telecommunications Capability to All Americans in a Reasonable and Timely Fashion, and possible Steps to Accelerate Such Deployment Pursuant to Section 706 of the Telecommunications Act of 1996," FCC 04-55, March 17, 2004, p. 6.

[57] FCC News Release, *FCC Improves Data Collection to Monitor Nationwide Broadband Rollout*, November 9, 2004. Available at [http://hraunfoss.fcc.gov/edocs_public/ attachmatch/DOC-254115A1.pdf]

[58] U.S. Government Accountability Office, Broadband Deployment is Extensive throughout the United States, but It Is Difficult to Assess the Extent of Deployment Gaps in Rural Areas, GAO-06-426, May 2006, p. 3.

[59] See Leighton, Wayne A., *Broadband Deployment and the Digital Divide: A Primer*, a Cato Institute Policy Analysis, No. 410, August 7, 2001, 34 pp. Available at [http://www.cato.org/pubs/pas/pa410.pdf]. Also see Thierer, Adam, *Broadband Tax*

Credits, the High-Tech Pork Barrel Begins, Cato Institute, July 13, 2001, available at [http://www.cato.org/tech/tk/010713-tk.html].

[60] See for example: Cooper, Mark, Consumer Federation of America and Consumers Union, *Expanding the Digital Divide and Falling Behind on Broadband*, October 2004, 33 pages. Available at [http://www.consumersunion.org/pub/ddnewbook.pdf]

[61] See CRS Report RL33542, *Broadband Internet Access: Background and Issues*, by Angele A. Gilroy and Lennard G. Kruger, for a discussion of regulatory issues.

In: Computer Networks, Policies and Applications
Editors: C. M. Berger and A. D. Streusse
ISBN: 978-1-61209-088-7

Chapter 3

IMPLEMENTATION AND EXPERIMENTATION OF A LAYER-2 ARCHITECTURE FOR INTERCONNECTING HETEROGENEOUS AD HOC NETWORKS TO THE INTERNET

Emilio Ancillotti[1,*], ***Raffaele Bruno***[2,†], ***Marco Conti***[2,‡], ***Enrico Gregori***[2,§] ***and Antonio Pinizzotto***[2,¶]
[1]Dept. of Information Engineering, University of Pisa
Via Diotisalvi 2 - 56122 Pisa, Italy
[2]IIT Institute, National Research Council (CNR)
Via G. Moruzzi, 1 - 56124 Pisa, Italy

Abstract

A prerequisite for the mass-market deployment of multi-hop ad hoc technologies is the capability of cooperating with existing wired infrastructure networks and providing transparent access to the Internet. However, current solutions to support connectivity between ad hoc networks and the Internet are based on complex mechanisms, such as Mobile-IP and IP tunnelling. In this chapter, we describe a lightweight and efficient architecture to build multi-hop hybrid ad hoc networks (i.e., using both wired and wireless links) employing proactive routing protocols, which can be used as flexible and low-cost extensions of traditional wired LANs. Our proposed architecture provides transparent global Internet connectivity and self-configuration to mobile nodes, without requiring configuration changes in the pre-existing wired LAN and relying only on basic layer-2 functionalities. We have prototyped the core components of this architecture for OLSR-based ad hoc networks. In this chapter, we present several experimental results such as to evaluate the network performance, and to determine an "optimal" setting of the routing protocol parameters.

[*]E-mail address: emilio.ancillotti@iet.unipi.it
[†]E-mail address: raffaele.bruno@iit.cnr.it
[‡]E-mail address: marco.conti@iit.cnr.it
[§]E-mail address: enrico.gregori@iit.cnr.it
[¶]E-mail address: antonio.pinizzotto@iit.cnr.it

1. Introduction

In spite of the massive efforts in researching and developing mobile ad hoc networks in the last decade, this type of networks has not yet witnessed a widespread deployment. The low commercial penetration of products based on ad hoc networking technologies could be explained by considering that users are interested in general-purpose applications where high bandwidth and open access to the Internet are consolidated and cheap commodities. For these reasons, there is a growing interest in designing working mechanisms for providing an easy access to the Internet to nodes in ad hoc networks.

A variety of architectural issues arise to offer Internet connectivity to mobile nodes in ad hoc networks. First of all, in the traditional view of ad hoc networks, no specific assumption is made about the IP addresses of mobile nodes. This means that each IP address serves only as a node's identifier and it does not convey any information about where the node is topologically located with respect to another node. As a consequence, routing in ad hoc networks is typically performed using only host specific routes. On the other hand, traditional IP routing is hierarchical and based on aggregated routes: one entry in the routing table can handle all hosts that share the same network identifier. Thus, specific mechanisms are needed to enable mobile nodes to distinguish between IP addresses belonging to the ad hoc components or the wired part of this extended WLAN. Another problem that arises when providing ad hoc nodes with connectivity to external networks is how to allow that the return traffic is correctly routed from the rest of the Internet to the ad hoc source node. In addition to these general issues, there are other problems more related to the management of node mobility in terms of roaming between the different ad hoc components. Some kind of signalling is needed to detect the different available gateways, to select one of them and to handoff to a new gateway when appropriate.

Two classes of approaches have been proposed in the literature to address the above mentioned issues in order to support connectivity between ad hoc networks and the Internet. One approach is to implement a Mobile IP Foreign Agent (MIP-FA) in the ad hoc node that acts as Internet gateway, and to run an extended implementation of Mobile IP in all the ad hoc nodes [1]. A different approach relies on the implementation of a Network Address Translation (NAT) in the gateway [2], which translates the IP addresses of ad hoc nodes to an address on the NAT gateway, which is routable on the external network. These approaches have been applied to both reactive and proactive ad hoc networks, and working solutions have been implemented and validated. However, these solutions relies on complex IP-based mechanisms originally defined for the wired Internet, like the IP-in-IP encapsulation, or explicit IP tunnelling [3], which introduce significant overheads.

This chapter describes an alternative approach to provide global Internet connectivity for reactive ad hoc networks, using basic layer-2 capabilities such as the ARP protocol [4]. In particular, our solution allows to build multi-hop hybrid ad hoc networks (i.e., using both wired and wireless links) employing proactive routing protocols, which can be used as flexible and low-cost extensions of traditional wired LANs. By positioning our architecture at the layer 2 (the data link layer), we may avoid undesired and complex interactions with the IP protocol and provide global Internet connectivity and node self-configuration in a very straightforward way. In addition, we also propose a dynamic protocol for the self-configuration of the ad hoc nodes, which relies on DHCP servers located in the wired part of the network, and it does not require that the ad hoc node to be configured has a direct access to the DHCP server. Using our scheme, mobile nodes can obtain a unique IP address that is topologically correct within the extended LAN.

We have prototyped the main components of our architecture in a general and realistic test-bed based on the OLSR routing protocol [5]. In this test-bed, we have conducted a large variety of tests in which we have carried out a large variety of tests, evaluating the throughput performance of Internet access under various network configurations and mobility conditions. The shown experimental results indicate that: *i*) the throughput performance of Internet access in static configurations can be significantly enhanced (in some cases we have more than doubled the measured throughputs) by properly setting the OLSR protocol parameters such as to improve route stability; and *ii*) the continuity of TCP sessions during node mobility is achieved without requiring additional overheads.

The rest of this chapter is organized as follows. Section 2. introduces our network model. In Section 3. we discuss available solutions for Internet connectivity and node self-configuration in MANETs. Section 4. outlines the relevant protocols used in our architecture. In Section 5. we describe the layer-2 architecture we propose to build hybrid ad hoc networks interconnected to the Internet. Section 6. shows experimental results on the performance constraints of our solution. Finally, Section 7. draws concluding remarks and discusses future work.

2. Network Model

In this chapter, we concentrate on investigating how the ad hoc networking paradigm can be applied to extend the range of traditional Wireless Local Area Networks (WLANs) [6] over multiple radio hops, in order to provide seamless and untethered mobility support and global Internet access for mobile/portables devices in the local area environment. Ensuring a seamless coverage when deploying traditional infrastructure-based WLANs is a difficult challenge to address, because of the several factors that impair the radio transmissions, such as electromagnetic interference, fading, obstacles, etc. Integrating ad hoc networking in traditional WLAN technologies allows discovering and maintaining multi-hop wireless paths within the network, providing a flexible, robust and cost-effective solution to increase coverage areas. More precisely, we envisage a hybrid network environment in which wired and multi-hop wireless technologies transparently coexist and interoperate. In this network, separated group of nodes without a direct access to the networking infrastructure form *ad hoc "islands"*, establishing multi-hop wireless links. Special nodes, hereafter indicated as *gateways*, having both wired and wireless interfaces, are used to build a wired backbone interconnecting separated ad hoc components. To ensure routing between these ad hoc parts, an ad hoc routing protocol is implemented on both gateways' interfaces. In addition, the gateways use their wired interfaces also to communicate with static hosts belonging to a wired LAN. The network resulting from the integration of the hybrid ad hoc network with the wired LAN is an *extended* LAN, in which static and mobile hosts transparently communicate using traditional wired technologies or ad hoc networking technologies. Note that a similar architecture has been adopted in a previous work dealing with connecting ad hoc networks to the Internet, such as [1, 7–9].

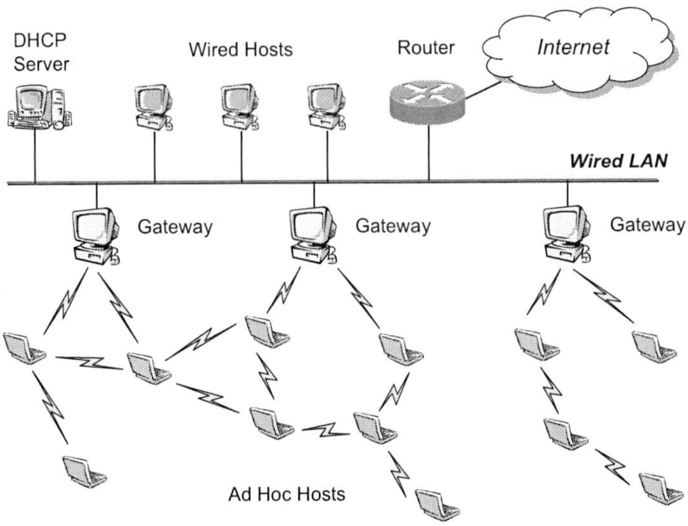

Figure 1. Reference Network Model.

Figure 1 illustrates the reference network model we assume in our architecture. We consider a full-IP network in which all the traffic is transported in IP packets. In this network, mobile/portable nodes far away from the fixed networking infrastructure establish multi-hop wireless links to communicate (e.g., using IEEE 802 .11 technology). As shown in the figure, gateways, i.e., nodes with two interfaces - both wired and wireless - are used to connect the ad hoc components to a wired LAN (e.g., an Ethernet-based LAN). In our architecture, it is allowed the multi-homing, i.e., the presence of multiple gateways within the same ad hoc component. Consequently, specific mechanisms are required to support the handoff between gateways without TCP-connection breaks. In general, between pairs of gateways in radio visibility of each other, two direct links can be established, both wired and wireless. However, in our model we assume that the gateways always use the wired link to communicate. The motivations behind this requirement will be clearly discussed in Section 5.. However, this is a quite reasonable assumption, since wired links have higher bandwidth than wireless links, and the routing protocol should assign them a lower link cost.

The wired LAN is interconnected to the external Internet through a default router *R*. In addition, one or more DHCP servers are located in the wired LAN to allocate network addresses to hosts. In the following sections, we will explain how these DHCP servers could be used to assign IP configuration parameters also to the ad hoc nodes. One of the basic assumptions of our architecture is the fact that all the hosts in the extended LAN share the same IP prefix. This implies that all the hosts in the external Internet consider the extended LAN as a single IP subnet. This is implicitly guaranteed by the fact that the same DHCP servers are used to assign the IP addresses to both wired nodes and ad hoc nodes. To indicate the network identifier we use the standard notation IP_S/L, in which IP_S indicates the network prefix, and L is the network mask length.

Standard IP routing is used to connect the extended LAN to the Internet. However, a specific ad hoc routing protocol is needed to allow multi-hop communications among the ad hoc nodes. In this work we decided to use a proactive routing protocol as the ad hoc routing algorithm (such as the Optimized Link State Routing (OLSR) protocol [5] or the Topology Dissemination Based on Reverse-Path Forwarding (TBRF) routing protocol [10]). The motivation behind this design choice is that proactive routing protocols usually support gateways, allowing these nodes to use special routing messages to set up default routes in the ad hoc network. Indeed, default routes are an efficient mechanism to forward traffic that does not have an IP destination locally known to the ad hoc network. In addition, proactive routing protocols, adopting classical link state approaches, build the complete network-topology knowledge in each ad hoc node. This topology information could significantly simplify the operations needed to acquire Internet connectivity. In this work, the reference ad hoc routing algorithm is OLSR, but our architecture is general and it is equally applicable to other proactive routing protocols.

3. Related Work

The implemented solutions to provide Internet connectivity in MANETs are mainly based on two different mechanisms.

One approach is to set up a Mobile IP Foreign Agent (MIP-FA) in the gateway and to run Mobile IP [11] in the MANET. In this way, the ad hoc node may register the foreign agent care-of-address with its Home Agent (HA). Whenever an ad hoc node MN wants to contact an external host X, it uses its home address (i.e., a static IP address belonging to its home network) as source address. As a consequence, the return traffic is routed to the home network through standard IP routing. The HA intercepts the traffic, encapsulating it using the care-of-address, and it tunnels the encapsulated packets to the FA. The FA removes the outer IP header and delivers the original packets to the visiting host MN. Different versions of this approach have been proposed and implemented for proactive [1] and reactive [8, 12] ad hoc networks. A variety of drawbacks characterize these solutions. The first one is that in order to allow Mobile IP and ad hoc networking to cooperate it is needed to introduce further complexities and sub-optimal operations in the implementations of both Mobile IP and ad hoc routing protocols, because the FA and the mobile node cannot be considered on the same link. Probably an integrated design of Mobile IP and ad hoc routing functionalities might be much more efficient, minimizing the overheads. In addition, Mobile IP was designed to handle mobility of devices in case of relatively infrequent mobility. Thus the overheads introduced to manage roaming and handoffs between foreign agents are a relevant issue in MANET environments, where handoff functions operating at the data link layer may be more suitable. Finally, when the technique of default routes is used to route the traffic from the mobile node to the closest gateway, the use of Mobile IP can easily lead to triangle routing. In fact, when the mobile node moves it can get closer to a gateway different from the one to which it is currently registered. As a consequence, the forward traffic leaves the ad hoc network through one gateway while the return traffic enters the ad hoc network through the MIP-FA gateway to which the mobile node is registered.

An alternative solution to interconnect MANETs to the Internet is to implement a Network Address Translation (NAT) [13] on the gateway. In this way, the gateway may translate the source IP address of outgoing packets from the ad hoc nodes with an address of the NAT gateway, which is routable on the external network. The return traffic is managed similarly, with the destination IP address (i.e., the NAT-gateway address) replaced with the IP address of the ad hoc node. NAT-based solutions have been designed for both proactive [2] and reactive [14] ad hoc networks. NAT-based mechanisms appear as easier solutions than MIP-FA-based schemes to provide Internet access to MANETs. However, a problem that arises with NAT-based solutions is multi-homing, i.e., the support of multiple gateways in the same MANET. Indeed, to avoid session breakages it is necessary to ensure that all the packets from the same session are routed over a specific gateway. A proposed solution to this issue is to explicitly tunnel all the outgoing traffic from the same communication session destined to the external network to one of the available gateways, instead of using default routes. A limitation of this strategy is the additional overhead introduced by the IP-in-IP encapsulation. Moreover, the ad hoc nodes should be provided with the additional capability of explicitly discovering the available gateways. This would eventually require extensions to the ad hoc routing protocols.

Both the two classes of solutions discusses above implicitly assume that either there is a dynamic host configuration protocol designed to configure the nodes such as to properly working in the MANET, or the ad hoc nodes are configured *a priori*. Indeed, a node in an IP-based network requires a unique IP-based address, a common netmask and, eventually, a default gateway. In traditional networks, hosts rely on centralized servers like DHCP [15] for configuration, but this cannot be easily extended to MANETs because of their distributed and dynamic nature. However, various protocols have been proposed recently in literature for the purpose of address self-configuration in MANETs. In general, with protocols using stateless approaches nodes arbitrarily select their own address, and a Duplicate Address Detection (DAD) procedure is executed to verify its uniqueness and resolve conflicts. On the other hand, protocols based on stateful approaches execute distributed algorithms to establish a consensus among all the nodes in the network on the new IP address, before assigning it. The protocols proposed in [16] and [17] are examples of the latter and former approach, respectively, while [18] presents a general overview of the several solutions currently available. Generally, all these protocols assume reliable flooding in order to synchronize nodes' operations and resolve inconsistencies in the MANET, but this is difficult to be guaranteed in ad hoc networks. Another main limitation of these solutions is that they are designed to work in *stand-alone* MANET, while no protocols have been devised to take fully advantage of the access to external networks. In addition, the problems of selecting a unique node address, routing the packets and accessing the Internet are still separately addressed, while a unified strategy may be beneficial, reducing complexities and overheads.

4. Protocol Descriptions

This section gives a short description of the protocols, which our architecture is based on.

4.1. OLSR

The OLSR protocol [5] is an optimization of the classical link state algorithm tailored to mobile ad hoc networks. More precisely, being a proactive routing protocol, OLSR periodically floods the network with route information, so that each node can locally build a routing table containing the complete information of routes to all the nodes in the ad hoc network running on their interfaces the OLSR protocol. The OLSR routing algorithm employs an efficient dissemination of the network topology information by selecting special nodes, the *multipoint relays* (MPRs), to forward broadcast messages during the flooding process. More precisely, each node independently selects its multipoint relays among its one-hop neighbors such as to ensure that a broadcast message, retransmitted by these selected neighbors, will be received by all its two-hop neighbors. The link state reports, which are generated periodically by MPRs, are called TOPOLOGY CONTROL (TC) messages. These TC messages are flooded to all the nodes in the network, but only the MPRs are allowed to forward the control messages received from other nodes, in order to reduce the number of retransmissions needed to cover the entire network.

Similarly to other routing protocols for ad hoc networks [19], OLSR employs a neighbor discovery procedure based on HELLO messages. These HELLO packets contain the list of neighbors known to the node and their link status. Thus, H ELLO messages allow each node to discover its one-hop neighbors, as well as its two-hop neighbors, which are needed during the MPR selection procedure. The neighborhood information and the topology information are updated periodically, and they enable each node to locally compute the least-cost routes to any possible destination in the ad hoc network, by using the Dijkstra's shortest path algorithm. This routing table is recomputed whenever there is a change in either the neighborhood information or the topology information.

In order to allow the injection of external routing information into the ad hoc network, the OLSR protocol defines the HOST AND NETWORK ASSOCIATION HNA message. The HNA message binds a set of network prefixes to the IP address of the node attached to the external networks, i.e., the gateway node. In this way, each ad hoc node is informed about the network address and netmask of the network that is reachable through each gateway. In other words, the OLSR protocol exploits the mechanism of *default routes* to advertise Internet connectivity. For instance, a gateway that advertises the 0.0.0.0/0 default route, will receive each packet destined to IP addresses without a known route on the local ad hoc network.

4.2. ARP Protocol

IP-based applications address a destination host using its IP address. On the other hand, on a physical network individual hosts are known only by their physical address, i.e., MAC address. The ARP protocol [20] is then used to translate, inside a physical network, an IP address into the related MAC address. More precisely, the ARP protocol broadcasts the *ARP_Request* message to all hosts attached to the same physical network. This packet contains the IP address the sender is interested in communicating with. The target host, recognizing that the IP address in the packet matches its own, returns its MAC address to the requester using an unicast *ARP_Reply* message. To avoid continuous requests, the hosts keep a cache of ARP responses.

In addition to these basic functionalities, the ARP protocol has been enhanced with more advanced features. For instance in [4] it has been proposed the *Proxy-ARP* mechanism, which allows constructing local subnets. Basically, the Proxy ARP technique allows one host to answer the ARP requests intended for another host. This mechanism is particularly useful when a router connects two different physical networks, say *NetA* and *NetB*, belonging to the same IP subnet. By enabling the Proxy ARP on the router's interface attached to *NetB*, any host A in *NetA* sending an ARP request for a host B in *NetB*, will receive as response the router's MAC address. In this way, when host A sends IP packets for host B, they arrive to the router, which will forward such packets to host B.

5. Proposed Architecture

Our design goal in the definition of the rules and operations of the proposed architecture is to provide transparent communications between static nodes (using traditional wired technologies) and mobile nodes (using ad hoc networking technologies), employing mechanisms that run below the IP layer. As discussed in the introduction, in this work we address two relevant issues: node self-configuration and global Internet connectivity.

5.1. Ad Hoc Node Self-configuration

The main obstacle to use a DHCP server for self-configuration of ad hoc nodes is that the DHCP server may be not reachable to the new node, due to mobility or channel impairments. In addition, the ad hoc nodes may need multi-hop communications to reach the DHCP server, but a unique address is necessary to execute ad hoc routing algorithms capable of establishing such communications. To solve these problems, we assume that the DHCP servers are located only in the wired part of the network, while in the ad hoc part of the network we implement dynamic *DHCP Relay* agents. These are special relay nodes passing DHCP messages between DHCP clients and DCHP servers that are on different networks.

Figure 2 shows what happen when a new node, say *MN*, is searching for a wireless connectivity service. We envision that a mobile node entering an area not covered by an infrastructure-based wireless LAN, starts a discovery procedure to find out if there is ad hoc connectivity. To execute this discovery procedure, the node *MN* broadcasts special messages, called NEIGHBOR_REQ, to find out other nodes that are within its radio visibility and that can interconnect him to the ad hoc component. To make more robust the protocol to messages' losses and various network configurations, node *MN* can periodically generate new NEIGHBOR_REQ messages scanning each channel and operating mode (e.g., 802.11*abg*) supported by its interface. As shown in Figure 2 more than one ad hoc node can be in radio visibility of node *MN*, such as to be able to receive this request and to reply. Note that the replies are sent in an unicast fashion to node *MN* because the NEIGHBOR_REQ messages convey the physical address of the requester. Node *N* may receive multiple replies, but it will elect just one of the responder as its DHCP relay agent according to some heuristic (e.g., the responder with the best signal quality, or the last one to reply). A unicast message can be used to inform the selected ad hoc node, for instance *H*2 in our example, that it can start a standard DHCP message exchange with the external DHCP server on behalf of node *MN*. The configuration process is concluded when the DHCP Relay forwards the DHCP_ACK message to the initial node *MN* that is now configured and can join the network. After joining the network, node *MN* may also turn itself into a DHCP relay agent for the DHCP server from which it received the IP configuration parameters, letting other

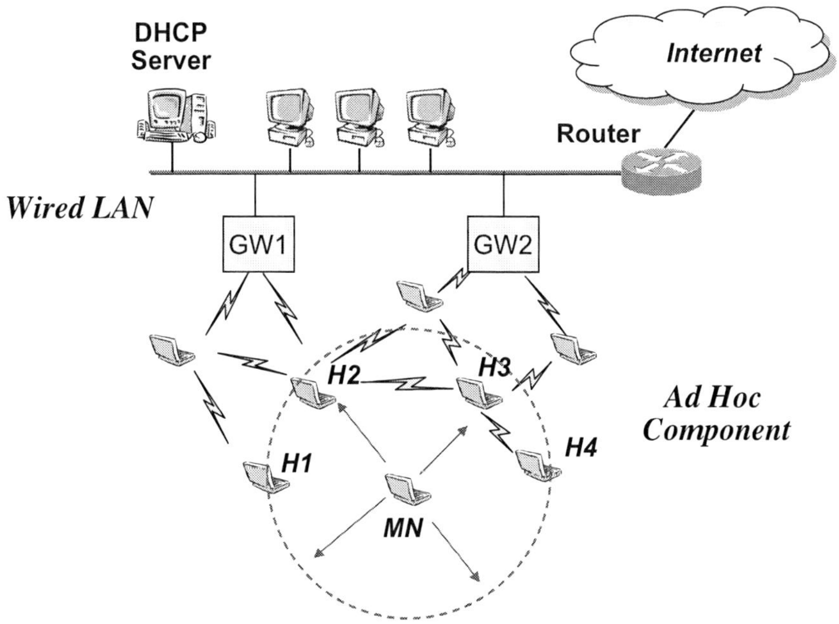

Figure 2. Relay agents discovery.

nodes to subsequently joining the ad hoc component. Finally, it is worth noting that it is not needed any initialization procedure for the ad hoc network, because the gateways are directly connected to the wired LAN and can broadcast a DHCP_DISCOVER message to locate available servers. In this way, the first mobile node to enter the ad hoc network may find at least one gateway capable of initiating the illustrated configuration process.

Our proposed node self-configuration mechanism is somehow similar to the one described in [17]. In that paper, a preliminary message handshake was used to discover a reachable MANET node that could act as initiator of the configuration process. On the contrary, in our solution the initiator node exploits the resources of the external wired network to which the ad hoc component is connected, to perform the IP address resolution.

5.2. Global Internet Connectivity

Our design goal is to support Intranet connectivity (i.e., communications with nodes inside the same IP subnet) and Internet connectivity (i.e., communications with nodes of external IP networks) for the mobile nodes, without any configuration change in the pre-existing wired LAN. The assumption that we take as starting point in our proposal is that the mobile nodes are configured with an IP address belonging to the same IP subnet of the wired LAN. This is achieved using the mechanism described in Section 5.1..

In the following we will separately explain how the proposed architecture ensures connectivity for outgoing and incoming traffic.

5.2.1. Connectivity for Outgoing Traffic

As outlined in Section 4.1., the OLSR protocol builds the routing tables with entries that specify the IP address of the next-hop neighbor to contact to send a packet destined to either another host or subnetwork. More precisely, to send a packet to a destination IP address, the mobile host searches for the longest IP prefix in the routing table matching the destination IP address. The matching routing table entry provides the next hop to send the packet. Since the gateways advertise $0.0.0.0/0$ as default route, all packets destined for IP addresses without a specific route on the ad hoc network, will be routed on a shortest-hop basis to the nearest gateway and forwarded to the Internet. However, using $0.0.0.0/0$ as default route for outgoing packets, introduces an inconsistency when a mobile host sends IP packets to a wired host inside the LAN. To explain this problem let

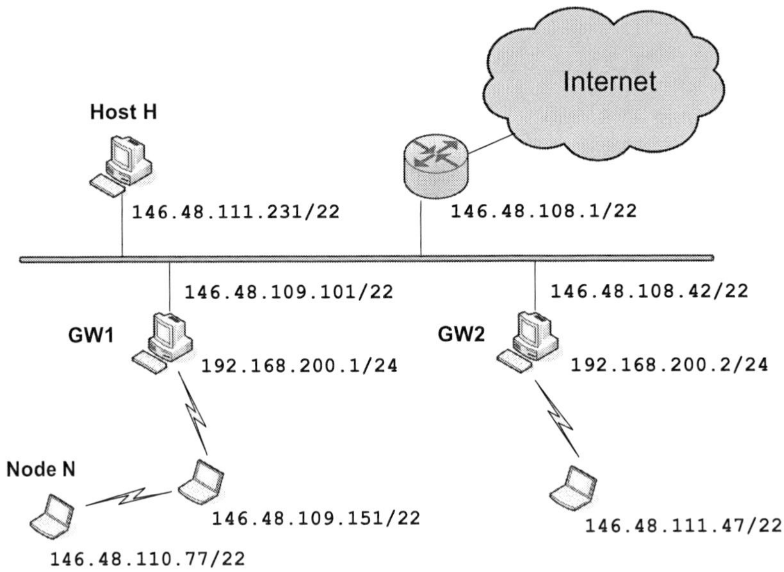

Figure 3. Illustrative network configuration.

us consider the simple network configuration depicted in Figure 3. For illustrative purposes we assume that the IP subnet of the extended LAN is $IP_S/L = 146.48.108.0/22$[1]. If the mobile node N ($IP_N = 146.48.110.77/22$) wants to deliver packets to the wired node H ($IP_H = 146.48.111.231/22$), the routing table lookup on node N will indicate that the node H is connected to the same physical network of node N's wireless interface. This will result in a failed ARP request for the IP_H address. To resolve this inconsistency, we will exploit the properties of the IP longest-matching rules. More precisely, we split the original IP subnet into two consecutive smaller subnets $IP_{SL}/(L+1)$ and $IP_{SU}/(L+1)$, such as to have that the union of these two sets is equal to IP_S/L. In the considered case

[1]On the gateways' wireless interfaces we set up private IP addresses to save address space. In this way, the gateways are globally reachable using the IP address on their wired interfaces.

$IP_{SL}/(L+1) = 146.48.108.0/23$ and $IP_{SU}/(L+1) = 146.48.110.0/23$[2]. Then, we configure all the gateways in such a way that they announce, through the HNA messages, also the connectivity to these two subnetworks. In this way, each mobile host will have, for any host on the local wired LAN, a routing table entry with a more specific network/mask than the one related to its wireless interface. To better clarify this point, let us consider the node N's routing table as shown in Table 1. The entries 7, 8, and 10 are the ones induced by the HNA messages arrived from GW1. The entry 9 is automatically set up by the operating system when the wireless interface is configured with the IP parameters. However, when searching the routing table for matching the IP_H address, node N will found the routing entry 8 more specific than entry 9. Consequently, the longest-match criterion applied to the routing table lookup, will result in node N correctly forwarding traffic to gateway GW1 (i.e., the nearest one) to reach node H.

Table 1. Node N's Routing Table.

Entry	*destination*	*next hop*	*metric*	*interface*
1	146.48.109.151/32	0.0.0.0	1	*eth*0
2	146.48.109.101/32	146.48.109.151	2	*eth*0
3	146.48.108.42/32	146.48.109.151	3	*eth*0
4	146.48.111.47/32	146.48.109.151	4	*eth*0
5	192.168.200.1/32	146.48.109.151	2	*eth*0
6	192.168.200.2/32	146.48.109.151	3	*eth*0
7	146.48.108.0/23	146.48.109.151	2	*eth*0
8	146.48.110.0/23	146.48.109.151	2	*eth*0
9	146.48.108.0/22	0.0.0.0	0	*eth*0
10	0.0.0.0/0	146.48.109.151	2	*eth*0
11	127.0.0.0/8	127.0.0.1	0	*lo*

The mechanism described above resolves any possible IP inconsistency that could occur in the mobile hosts, but it may cause problems for the gateways. In fact, being part of the ad hoc component, the gateways will receive HNA messages sent by other gateways, setting up the additional routing entries advertised in these messages. However, when a gateway wants to send packets to a wired host on the local wired LAN (e.g., node H), the routing table lookup will choose one of these two entries, instead of the entry related to its wired interface (i.e., 146.48.108.0/22). The effect is that the IP packet will loop among the GW nodes until the TTL expires, without reaching the correct destination H. To resolve this problem, we statically add in each gateway two further routing entries in addition to the one related to the default router 146.48.108.1. These two additional entries have the same network/mask as the two announced in the HNA messages, but with lower metric. Again, to better clarify the routing operations, let us consider the illustrative example shown in Figure 3. In Table 2 we have reported the GW1's routing table. In this example, *eth*0 is

[2]It is straightforward to observe that this operation is always feasible, at least for $L < 32$.

the GW1's wireless interface and *eth*1 is the GW1's wired interface. When gateway GW1 wants to send packets to node H, it will found two routing table entries matching the same number of bits of node H's IP address. These are entry 8 (derived from HNA messages received from GW2) and entry 10 (statically configured on the gateway). However, entry 10 has a lower metric than entry 8 (i.e., metric 0 against metric 1). As a consequence, the packets destined to host H can be correctly forwarded to the host H on the local wired LAN through the GW1's wired interface.

Table 2. GW1's Routing Table.

Entry	*destination*	*next hop*	*metric*	*interface*
1	146.48.109.151/32	0.0.0.0	1	*eth*0
2	146.48.110.77/32	146.48.109.151	2	*eth*0
3	146.48.108.42/32	0.0.0.0	1	*eth*1
4	146.48.111.47/32	146.48.108.42	2	*eth*1
5	192.168.200.2/32	146.48.108.42	1	*eth*1
6	192.168.200.0/24	0.0.0.0	0	*eth*0
7	146.48.108.0/23	146.48.108.42	1	*eth*1
8	146.48.110.0/23	146.48.108.42	1	*eth*1
9	146.48.108.0/23	0.0.0.0	0	*eth*1
10	146.48.110.0/23	0.0.0.0	0	*eth*1
11	146.48.108.0/22	0.0.0.0	0	*eth*1
12	0.0.0.0/0	146.48.108.1	0	*eth*1
13	0.0.0.0/0	146.48.108.42	1	*eth*1
14	127.0.0.0/8	127.0.0.1	0	*l*0

5.2.2. Connectivity for Incoming Traffic

A mechanism is required to ensure that the return traffic coming from hosts on the local wired LAN or from the Internet (through the default LAN router, as shown in Figure 1), gets correctly routed to the mobile hosts. Our basic idea is to introduce specific Proxy ARP functionalities into each gateway, in such a way that the gateways can hide the ad-hoc node identity on the wired physical network, which the gateways are connected to. Thus, all mobile nodes located in the ad hoc component will appear to wired hosts as being one IP-hop away. Internally to the ad hoc component, the ad hoc routing protocol will transparently provide the multi-hop connectivity and the mobility support. This is somehow similar to what is implemented in the LUNAR framework [21], in which the entire ad hoc network appears as a single virtual Ethernet interface.

In our proposed solution, a Proxy ARP server runs on the wired interfaces of each gateway. The Proxy ARP server periodically checks the gateway's routing table and ARP table, such as to publish the MAC address of the gateway's wired interface for each IP address having an entry in the routing table with a netmask 255 .255.255.255, and the next hop on the gateway's wireless interface. The former condition is verified only by routing table entries targeting mobile hosts that have joined the ad hoc network. The latter condition implies that the gateway can deliver traffic to that node only over multi-hop paths not traversing other gateways[3]. Thus, it is highly probable that the considered gateway is the default gateway selected by that ad hoc node. To illustrate how the proposed mechanism works, let us consider the network in Figure 3. When a node on the wired local LAN (e.g., node H) wants to send packets to an ad hoc node (e.g., node N), it assumes that the ad hoc node is on the same physical network. Hence, node H checks its ARP table for IP-MAC mapping and, if it is not present, it sends an ARP request. The gateway GW1 fulfills the previously defined conditions (i.e., node N's IP address has an entry in the GW1's routing table with a netmask 255.255.255.255, which is related to its wireless interface), while GW2 does not. Consequently, only GW1 is allowed by the Proxy ARP server to answer with an ARP reply. This ARP reply will insert the mapping [node N's IP address - MAC address of GW1's wired interface] into the node H's ARP table. Thus, the packets sent from node H to node N will be delivered to GW1, which will forward them to node N. On the other hand, node N will reply to node H using GW1, as indicated by its routing table (see Table 1).

There are some network configurations where asymmetric routing may occur, i.e., the forward path is different from the return path. For instance, let us consider the case in which node N is in radio visibility of two gateways GW1 and GW2. In this situation, the OLSR routing algorithm will randomly select one of these gateways as default gateway for node N. However, both gateways are allowed to send ARP replies for ARP requests issued by node H for the node N's IP address. In this case, the wired node H will update its ARP table using the information delivered in the last received ARP reply. Let us assume that GW1 is the default gateway for node N, but GW2 has sent the last ARP reply to node H. In this case, node H sends the traffic destined to node N to GW2, which routes it to node N. On the other hand, node N sends packets destined to node H to GW1, which forwards them to node H. It is important to note that asymmetric paths are not by themselves a problem. Indeed, both node N and H correctly receive and send their packets. In addition the asymmetric routing occurs only in symmetric topologies. Thus, it is reasonable to assume, in this local environment, that both paths are characterized by similar delays.

5.2.3. Mobility Support

In general, solutions to support Internet connectivity for ad hoc networks, which are based on gateways, experience TCP-session breaks when the default route changes, depending on dynamics and mobility in the network. To avoid that TCP sessions break, in [14] it was proposed to replace default routes with explicit tunneling between the mobile nodes and the gateways. However, this complicates significantly the implementation and introduces relevant overheads. On the contrary, in our architecture the mobility is supported in a transparent way for the higher protocol layers. Indeed, the only effect of changing the

[3]It is worth reminding that gateways are always interconnected using their wired interfaces. Hence, a route to reach a mobile node can traverse two gateways only if one of the link along the path is a wired link. In this case the farthest gateway will have the next-hop routing entry for that mobile node on its wired interface.

default gateway for node N, is that the node N's outgoing traffic is routed towards the new gateway (e.g., GW2), while the initial gateway (e.g., GW1) continues to receive the incoming traffic and to forward it to node N. This results into asymmetric routing. However, this asymmetry can be easily removed by using an advanced feature of the ARP protocol. More precisely, when GW2 becomes aware that the next hop for the node N switches from its wired interface to its wireless interface, it generates a *Gratuitous ARP* on the wired interface for node N's IP address. This will update the ARP table in all of the wired hosts that have an old entry for the node N's IP address, which was mapped with the MAC address of GW1's wired interface. This action restores a symmetric path for the active packet flows destined to and/or originated from node N.

6. Experimental Results

We have prototyped the core functionalities of our architecture. In particular, we have developed the software components described in Section 5.2., concerning the support of Internet and Intranet connectivity for the ad hoc nodes. Currently, we are completing the implementation of the modifications to the DHCP Relay agents described in Section 5.1.. For these reasons, in the following we will show experimental results measuring the network performance with mobility and Internet access, while we left for further work the testing of the performance (such as address allocation latency and communication overheads) of the proposed node self-configuration scheme.

Our test-beds consists of a number of *IBM R-50* laptops with *Intel Pro-Wireless 2200* as integrated wireless card. We have used the OLSR_UniK implementation for Linux in version 0.4.8 [22]. The installed Linux kernel distribution is 2.6.9. The ad hoc nodes are connected via IEEE 802.11*b* wireless links, transmitting at the maximum rate of 11 Mbps. To generate the asymptotic UDP and TCP traffic during the experiments we have used the *iperf* tool[4]. More precisely, the iperf server (termination of the traffic sessions) runs in a static host in the wired LAN, while iperf clients (originators of traffic sessions) are set up on the mobile nodes. If not otherwise specified, the packet size is constant in all the experiments and the transport layer payload is equal to 1448 bytes. Differently from other studies [2], in which the network topology was only emulated by using the *IP-tables* feature of Linux, our experiments are conducted in realistic scenarios, with hosts located at the ground floor of the CNR building.

6.1. Performance Constraints of Internet Access

To measure the performance constraints in case of Internet access, we executed several experiments in the test-bed shown in Figure 4. The distances between the ad hoc nodes were set up in such a way to form a 4-hop chain topology with non-volatile wireless links.

The first set of experiments was conducted to evaluate the impact on the UDP and TCP throughput of the number of wireless hops traversed in the ad hoc network to reach the gateway. During these tests all the OLSR configuration parameters have been set up according to the default values indicated in the RFC specification [5]. The measurements

[4]http://dast.nlanr.net/Projects/Iperf/.

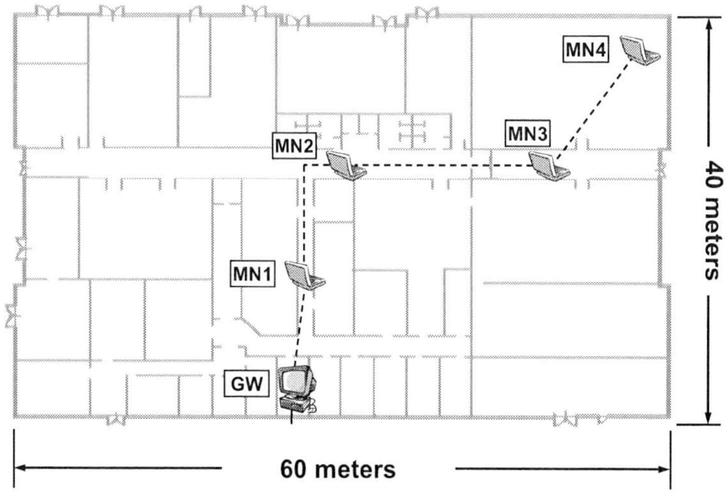

Figure 4. Trial scenario for testing Internet access using a chain network.

results are plotted in Figures 5. In particular, each graph compares the throughput obtained per unit time by a single UDP or TCP flow, varying the number of hops needed to reach the gateway. Several useful observations can be drawn form these measurements. First of all, the UDP throughput is always greater than the TCP throughput. This is an expected result, considering that the TCP protocol introduces the additional overhead of the TCP ACK packets. However, an unexpected outcomes of our experiments that is clearly pointed by the curves shown in Figures 5, is that both UDP and TCP flows are frequently blocked. In addition, the longer is the chain length in terms of hops, and the more frequent are the holes in the throughput performance.

To explain these phenomena we carried out an extensive investigation of traffic traces considering both the transport layer traffic and the routing traffic. We discovered that the throughput degradations are mainly due to the fact that links between neighboring nodes are frequently purged from the routing tables, resulting into a significant route instability. An initial explanation of this observed route instability is that losses of routing control messages can cause the timeout of valid routes. Indeed, the routing messages are delivered using broadcast frames, which are neither acknowledged nor retransmitted, hence they are more vulnerable to collisions and channel errors than unicast frames. However, a careful analysis of the routing log files has pointed out another relevant condition that contributes to the route instability in our static network. Indeed, we discovered that the OLSR protocol implements an over pessimistic estimation of the link quality that may cause the purge of heavy-loaded links. This process is called *link hysteresis* [5].

The link hysteresis is a procedure designed to make more robust the link sensing against bursty loss or transient connectivity between nodes. More precisely, the link hysteresis computes for each link between a node and its neighbors a value, called *link_quality*, which measures the reliability of that link. A link is considered "bad" if it allows H ELLO messages to pass through it sometimes but not very often. An established link, i.e., a symmetric and reliable link, is considered pending and not used for communications when its

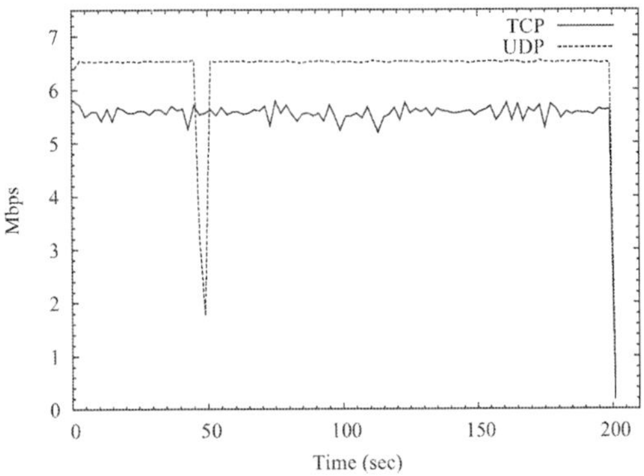

[1 hop]

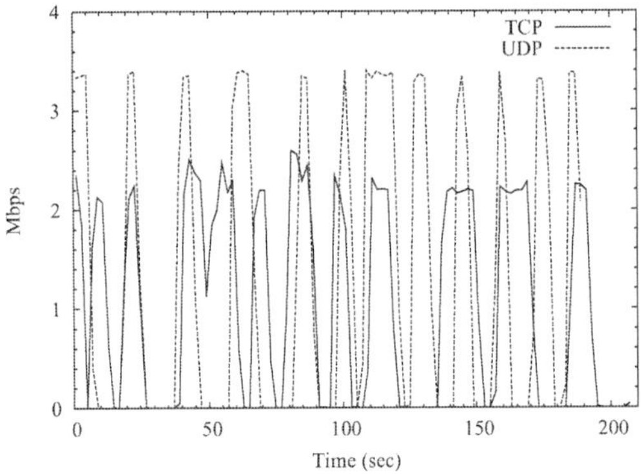

[2 hops]

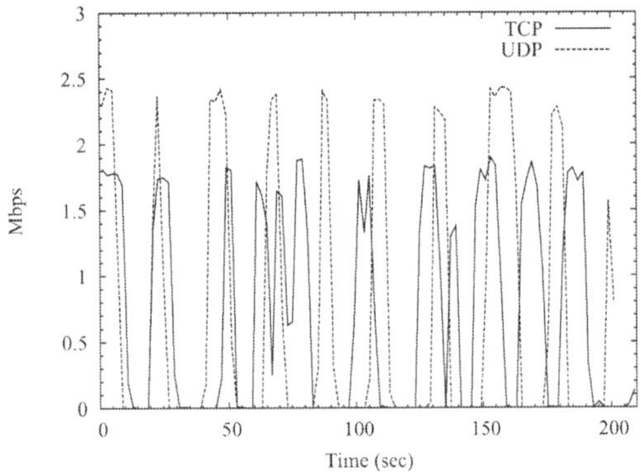

[3 hops]

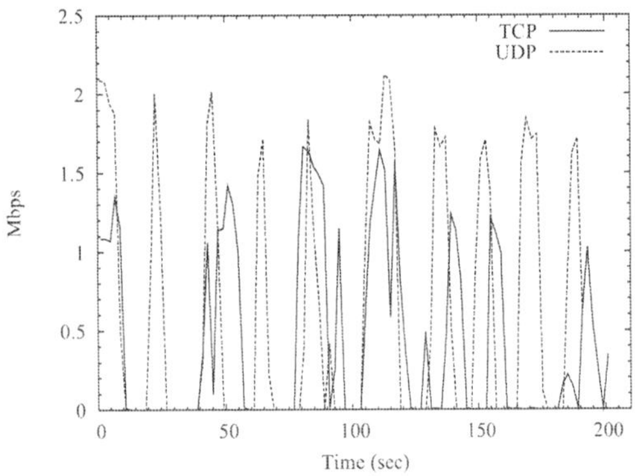

[4 hops]

Figure 5. Throughput of UDP and TCP flows for different chain lengths.

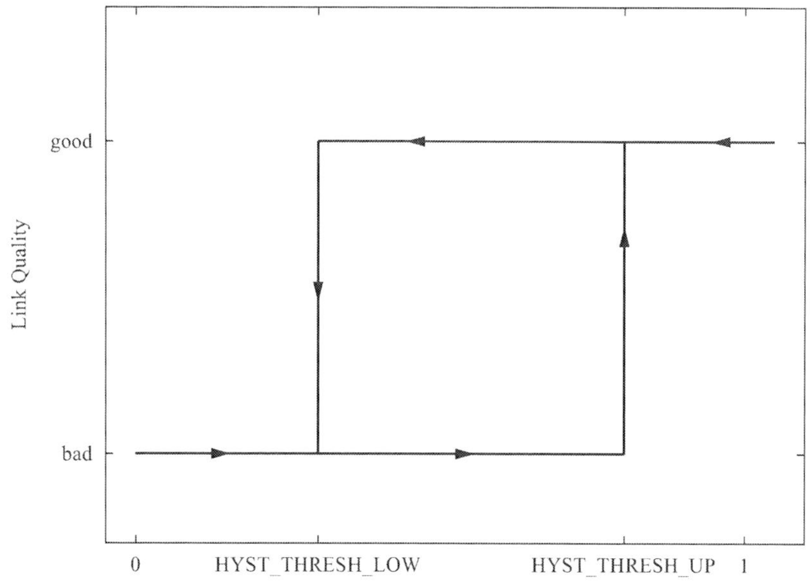

Figure 6. Illustration of the hysteresis process.

$link\_quality$ goes below a fixed threshold, known as $HYST\_THRESHOLD\_LOW$. Note that a pending link is non considered as lost because the link properties are still updated for each received HELLO message. On the contrary, a link is considered as lost when its validity time expires. In this case, the link is purged from the neighborood list. On the other hand, a pending link is promoted to the established status only when its $link\_quality$ goes above a fixed threshold known as $HYST\_THRESHOLD\_UP$. It is quite obvious that $HYST\_THRESHOLD\_UP \geq HYST\_THRESHOLD\_LOW$. Figure 6 illustrates the link hysteresis behavior. The diagram points our that when $HYST\_THRESHOLD\_LOW < link\_quality < HYST\_THRESHOLD\_UP$ the link status remains unchanged.

A key implementation requirement for the link hysteresis is the availability of an appropriate measure of the $link\_quality$. If some measure of the signal/noise level on a received message is available (e.g., as a link layer notification), then it can be used as estimation of the $link\_quality$ index (after being normalized in the range $[0,1]$). In [5], it is also described an algorithm to estimate the $link\_quality$ when it is impossible to use link-layer mechanisms. This algorithm monitors the loss of any OLSR packet, and it adopts an exponentially smoothed moving average of the OLSR-packet transmission success rate to measure the $link\_quality$. Formally, every time an OLSR packet is correctly received $link\_quality = (1 - HYST\_SCALING) \cdot link\_quality + HYST\_SCALING$, where the $HYST\_SCALING$ value is the smoothing factor of the estimator, which is a number fixed between 0 and 1. When an OLSR packet is lost, the instability rule is applied, that is $link\_quality = (1 - HYST\_SCALING) \cdot link\_quality$. Finally, a new link is initially set as pending and with a $link\_quality$ equal to $HYST\_SCALING$.

The behavior of the hysteresis strategy is clearly determined by the specific setting of the algorithm parameters, and in particular by the memory size of the $link\_quality$ estimator and the threshold values. The OLSR specification suggests as default configuration $HYST\_THRESHOLD\_LOW = 0.3$ and $HYST\_THRESHOLD\_UP = 0.8$, and it adopts $HYST\_SCALING = 0.5$ as scaling factor. According to these values, even a perfect link (i.e., a link with $link\_quality = 1$) will be purged from the routing tables when two consecutive OLSR packets are lost. We argue that the standard setting of the hysteresis parameters introduces a critical instability in the routing tables, because it is not infrequent to loose broadcast packets (as the OLSR packets are) when the channel is overloaded.

Table 3. UDP Throughput in a Chain Network, With and Without Hysteresis.

	UDP		
	HYST	NO HYST	
1 *hop*	6.124Mbps (304Kbps)	6.363Mbps (393Kbps)	+4%
2 *hops*	1.252Mbps (55Kbps)	2.501Mbps (57Kbps)	+100%
3 *hops*	700.4Kbps (60Kbps)	1.306Mbps (87Kbps)	+86%
4 *hops*	520.6Kbps (54Kbps)	1.141Mbps (56Kbps)	+119%

Table 4. TCP Throughput in a Chain Network, With and Without Hysteresis.

	TCP		
	HYST	NO HYST	
1 *hop*	5.184Mbps (335Kbps)	5.172Mbps (393Kbps)	$\approx=$
2 *hops*	956.1Kbps (123Kbps)	1.517Mbps (57Kbps)	+58%
3 *hops*	638.1Kbps (149Kbps)	891.7Kbps (77Kbps)	+39%
4 *hops*	345.9Kbps (47Kbps)	631.2Kbps (74Kbps)	+82%

To verify our claim we have carried out a second set of experiments in the same network configuration depicted in Figure 4, disabling the OLSR hysteresis process. To provide statistically correct results, we have replicated each experiment five times. Tables 3 and Table 4 show the average and standard deviation (in parenthesis) values of the measured throughputs for the UDP and TCP case, respectively. From the results we observe that the throughput performances are significantly improved, with the improvement for a 4-hop chain reaching 119% in the UDP case and 82% in the TCP case. The study of routing table logs clearly indicates that these throughput increases are due to an improvement in the route stability with less frequent declarations of link drops due to erroneous estimations of links' reliability. It is worth pointing out that this issue has not been identified in previous experimental studies because either the multi-hop communications were only emulated [14], or the channel was loaded with low-intensity *ping* traffic [23].

In addition to the hysteresis process, the OLSR protocol employs several other mechanisms, as the link sensing, neighbor detection and topology discovery, which significantly affect the route stability. Indeed, recent works [23, 24] have investigated how the setting of the classical OLSR routing parameters may affect the network performances. However, these works have specifically focused on the time required for route recalculation after a link drop due to node mobility. On the contrary, to conclude this section, we will analyze the impact of different OLSR parameter settings on the performance limits of Internet access in static network configurations. More precisely, each OLSR packet, and the information it delivers, has a fixed validity time. For instance, the information provided in a HELLO message is considered valid for a *NEIGHB_HOLD_TIME*. This implies that a node detects a link loss with a neighbor from the lack of HELLO messages during a *NEIGHB_HOLD_TIME*. A similar check is performed for the *TC* messages, whose validity time is *TOP_HOLD_TIME*, and for the HNA messages, whose validity time is *HNA_HOLD_TIME*. A possible strategy to avoid that links and routes are dropped from the routing tables because the related information has not been refreshed within the corresponding timeout, is to increase the frequency used to generate OLSR packets. This may increase the probability that at least one new OLSR packet is received before its validity time expires. The drawback of this approach is that the more frequent the OLSR protocol generates control messages, the higher is the routing overheads. To quantify the trade-off between routing overhead increases and route stability improvements, and how this impacts network performance, we have carried out a set of experiments in a 3-hop chain using the OLSR parameter settings shown in Table 5. As listed in the table, we compare the default parameter setting with disabled hysteresis (*set1*) with the cases in which the frequency of OLSR packet generations is two times (*set2*) and four times (*set3*) higher, while the validity times are kept constant.

Table 5. OLSR parameter configurations.

OLSR parameters	*set1*	*set2*	*set3*	*default*
HELLO_INTERVAL(s)	2	1	0.5	2
NEIGHB_HOLD_TIME(s)	6	6	6	6
TC_INTERVAL(s)	5	2.5	1.25	5
TOP_HOLD_TIME(s)	15	15	15	15
HNA_INTERVAL(s)	5	2.5	1.25	5
HNA_HOLD_TIME(s)	15	15	15	15
Hysteresis	no	no	no	yes

The experimental results obtained by replicating five times the throughout measurements for UDP and TCP traffic are listed in Table 6, in which the average throughput and its standard deviation (in parenthesis) are reported. The shown results indicate that increasing the frequency the OLSR packets are generated by a factor of four, and maintaining the default validity times, it is possible to improve the average throughput of 40% in the UDP case, and of 55% in the TCP case. To better explain the reasons of such an improvement, in Figure 7 and Figure 8 we show the UDP and TCP throughput, respectively, obtained during a single experiment, as a function of the time and the OLSR parameters settings. This results were obtained using a 3-hop wireless path to reach the gateway. Similarly to the behavior shown in Figures 5, we note that, although the nodes are static, the throughput

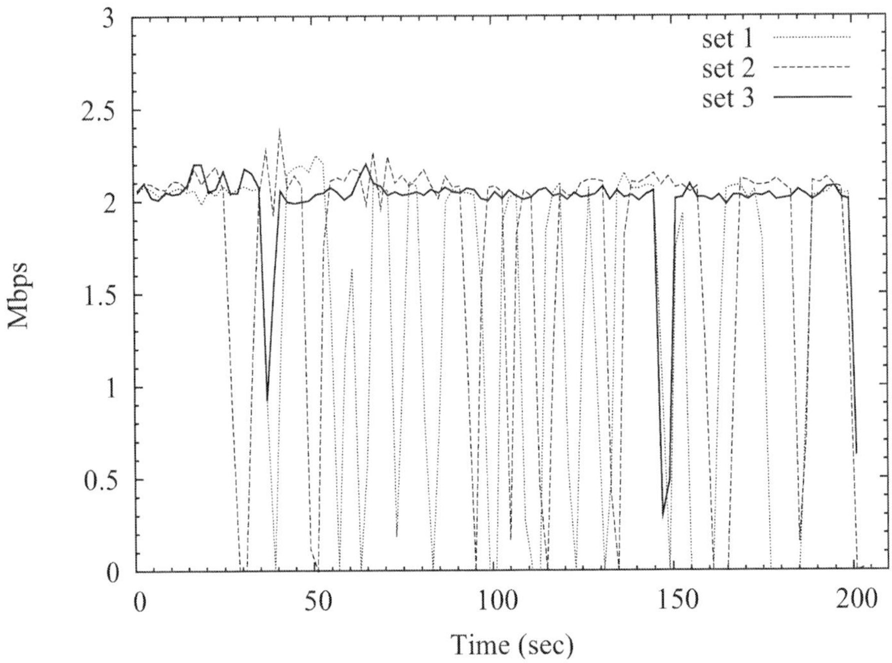

Figure 7. Impact of OLSR parameter setting on a 3-hop UDP flow.

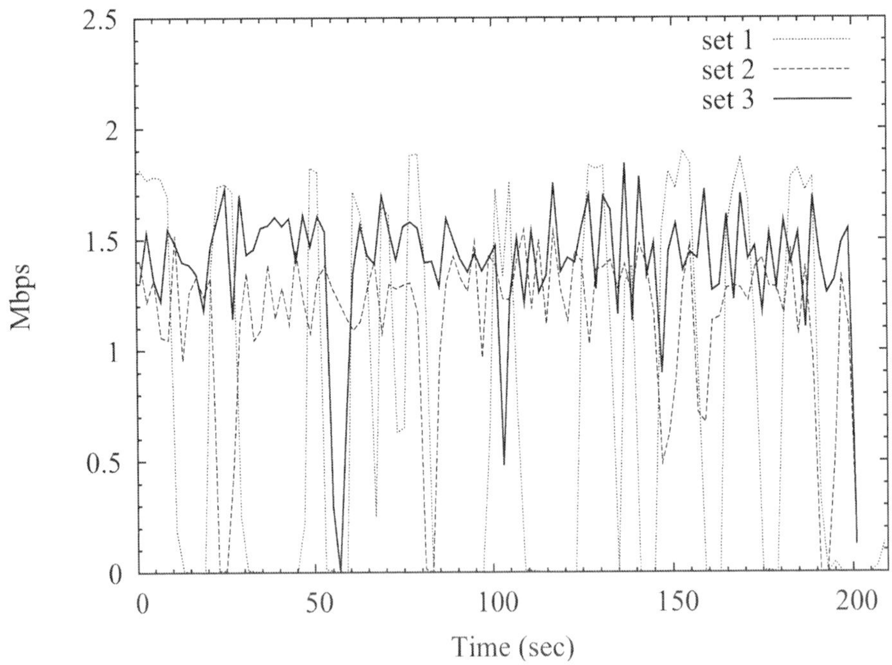

Figure 8. Impact of OLSR parameter setting on a 3-hop TCP flow.

Table 6. UDP and TCP Throughputs in a 3-hop Chain Network for Different OLSR Parameter Settings.

Parameter Setting	UDP	TCP
default	700.4Kbps (60Kbps)	638.1Kbps (149Kbps)
set1	1.306Mbps (87Kbps)	838.5Kbps (79Kbps)
set2	1.605Mbps (76Kbps)	1.020Mbps (105Kbps)
set3	1.84Mbps (106Kbps)	1.306Mbps (56Kbps)

is not stable, but both UDP and TCP flows could be in a stalled condition. However, the more frequent the OLSR routing packets are generated, the shorter the stalled conditions are. In summary, our experimental study indicates that the network performance of Internet access in static configurations can be significantly enhanced (in some cases we have more than doubled the measured throughputs) by properly setting the OLSR parameters such as to improve route stability.

By analyzing the routing table logs generated during the trials, we have discovered that the throughput increases are due to an improvement in route stability. On the other hand, the increase of routing overheads has a negligible impact on the throughput performance, because the peak throughout is almost independent of the adopted OLSR parameter setting.

In summary, our experimental study indicates that the network performance of Internet access in static configurations can be significantly enhanced (in some cases we have more than doubled the measured throughputs) by properly setting the OLSR parameters such as to improve route stability.

6.2. Performance Constraints with Mobility

To test the mobility support in a multi-homed network configuration we considered the network layout illustrated in Figure 9. In our experiments, node MN2 alternates between position P1 and position P2. More precisely, it starts in position P1, where it is in radio visibility of node MN1. After 50 seconds it moves in position P2, where it is in radio visibility of node MN3. The time needed for moving from P1 to P2 is 20 seconds. After other 50 seconds, host MN2 goes back to position P1. This mobility patterns is periodically repeated throughout the test. The HNA messages from GW1 and GW2 form default routes to the external network on a short-hop basis. Hence, while connected to MN1, the node MN2 uses GW1 as default gateway. On the contrary, when connected to node MN3, the routes are recalculated and MN2 uses GW2 as default gateway. The new default gateway GW2 will also begin to act as Proxy ARP for the mobile node. The return traffic will be consistently routed through the new gateway as soon as either a new ARP request for the MN2's IP address is issued by the external host, or the gateway GW2 sends a Gratuitous ARP; otherwise it will continue to arrive at the GW1 (see Section 5.2.3. for the details).

Figure 10 shows the TCP throughput achieved by MN2 during a mobility test. We compare these results against the throughput measured when node MN2 is fixed in position P1.

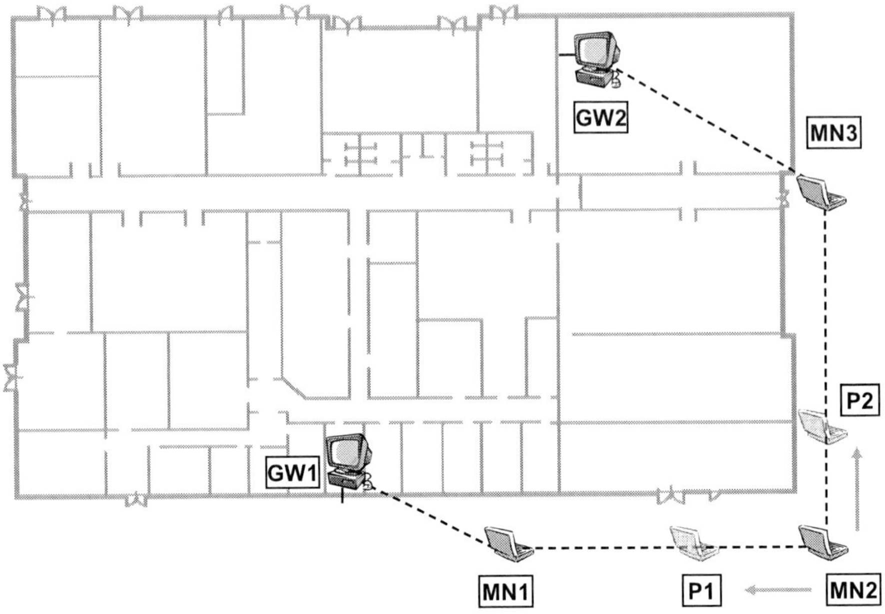

Figure 9. Trial scenario for testing mobility support.

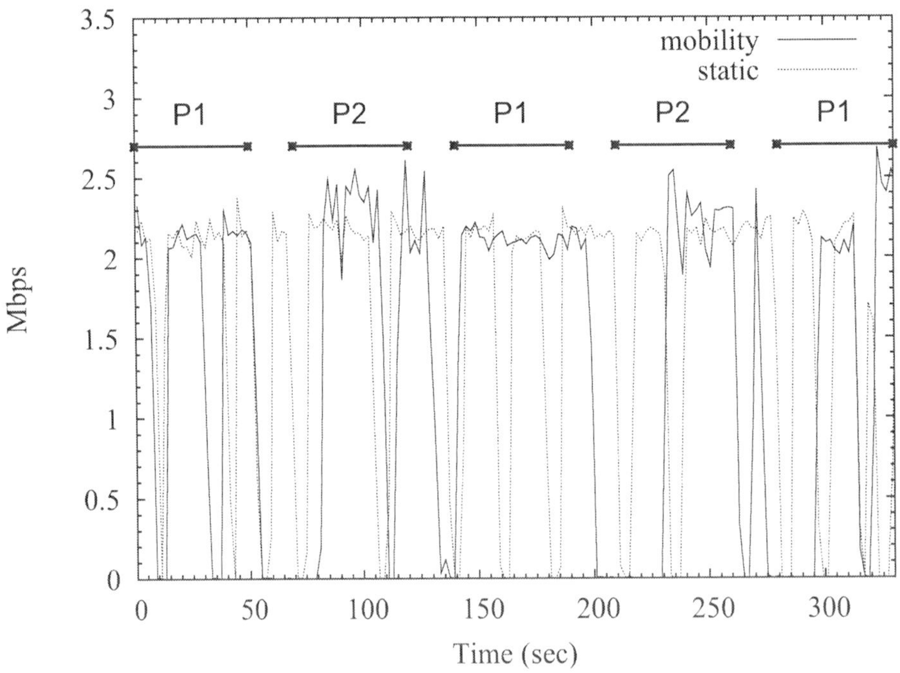

Figure 10. Throughput of a single TCP flow with node mobility.

During both experiments, the hysteresis process was disabled and the other OLSR parameters were set up according to the default values indicated in the RFC specification [5]. The shown results confirm that the TCP session does not break when node moves. The major effect of node mobility is to introduce holes in the TCP traffic due to the time needed to recalculate the new routes to reach the default gateway. In the considered case of "soft" handoff, i.e., the mobile nodes is in radio visibility of both node MN1 and node MN3 when changing position, we measured up to 20 seconds for recomputing a consistent routing table in node MN2. It is worth noting that in similar experiments conducted in [2], the throughput of mobile node was approximately 30% lower when mobile node changed position. This was expected because the TCP-session continuity was ensured at the cost of using IP tunneling that introduces significant additional overheads. On the contrary our solution is very efficient and lightweight, because it operates directly at the data link layer.

7. Concluding Remarks

This chapter presented a practical architecture to logically extend traditional wired LANs using multi-hop ad hoc networking technologies. Our design goal was to implement a set of mechanisms operating at the data link layer to interconnect a hybrid ad hoc network, formed by a mix of fixed and mobile nodes using both wired and multi-hop wireless technologies, to the Internet. Our architecture provides ad hoc node self-configuration and both Intranet and Internet connectivity in a way that is transparent to the wired nodes, i.e., without requiring changes in the pre-existing wired LAN. In addition, by locating our architecture below the IP level, we have designed a lightweight and efficient ad hoc support framework, which is easy to be implemented and introduces minimal overheads. We have prototyped the proposed architecture to test its functionalities. The shown experimental results indicate that: *i*) the network performance of Internet access in static configurations can be significantly enhanced (in some cases we have more than doubled the measured throughputs) by properly setting the OLSR protocol parameters such as to improve route stability; and *ii*) the continuity of TCP sessions during node mobility is achieved without requiring additional overheads.

We believe that there are several related aspects that are worth being further investigated in future work.

- The gateway selection procedure implicitly relies on the ad hoc routing protocol. In the case of OLSR, it is accomplished according to a shortest-path basis. However, in a multi-homing scenario, several gateways can exist, which may be implemented using different technologies and may have different capabilities. Thus, there could be many benefits in designing cooperative heuristics to select gateways such as to obtain load balancing within the ad hoc network, or more efficient handovers.

- In this work we have considered basic IP services, i.e., unicast routing and dynamic address allocation. However, more sophisticated functionalities, such as multicast and QoS management, have been developed for the Internet. Therefore, the proposed architecture should be extended to integrate these additional capabilities.

- The address allocation scheme described in this chapter allows the exploitation of

DHCP servers to assign IP addresses that are topologically correct in the entire extended LAN. However, there is not a detailed evaluation of efficiency of this proposal or a comparative analysis with different node auto-configuration schemes. In addition, it should be explored how to extend our solution to deal with the typical problems that may arise due to mobility of nodes, such as message losses, and network partition and merger.

References

[1] M. Benzaid, P. Minet, K. Al Agha, C. Adjih, and G. Allard, "Integration of Mobile-IP and OLSR for a Universal Mobility," *Wireless Networks*, vol. 10, no. 4, pp. 377–388, July 2004.

[2] P. Engelstad, A. Tønnesen, A. Hafslund, and G. Egeland, "Internet Connectivity for Multi-Homed Proactive Ad Hoc Networks," in *Proc. of IEEE ICC'2004*, vol. 7, Paris, France, June, 20–24 2004, pp. 4050–4056.

[3] C. Perkins, "IP Encapsulation within IP," RFC 2003, October 1996. [Online]. Available: http://www.ietf.org/rfc/rfc2003.txt

[4] D. Plummer, "An Ethernet Address Resolution Protocol," RFC 826, November 1982. [Online]. Available: http://www.ietf.org/rfc/rfc0826.txt

[5] T. Clausen and P. Jaquet, "Optimized Link State Routing Protocol (OLSR)," RFC 3626, October 2003. [Online]. Available: http://www.ietf.org/rfc/rfc3626.txt

[6] *Wireless LAN Medium Access Control (MAC)and Physical Layer (PHY) Specification/Amendment 2: Higher-speed Physical Layer (PHY) in the* 2.4 *GHz band*, The Institute of Electrical and Electronics Engineer Std. 802.11b, Nov. 2001.

[7] J. Broch, D. Maltz, and D. Johnson, "Supporting hierarchy and heterogenous interfaces in multi-hop wireless ad hoc networks," in *Proc. of I-SPAN'99*, Perth, Australia, June, 23–25 199, pp. 370–375.

[8] Y. Sun, E. Belding-Royer, and C. Perkins, "Internet Connectivity for Ad hoc Mobile Networks," *International Journal of Wireless Information Networks* , vol. 9, no. 2, pp. 75–88, April 2002, Special issue on "Mobile Ad hoc Networks: Standards, Research, Application".

[9] R. Bruno, M. Conti, E. Gregori, A. Pinizzotto, and E. Ancillotti, "Experimenting a Layer 2-based Approach to Internet Connectivity for Ad Hoc Networks," in *Proc. of IEEE ICPS REALMAN Workshop 2005*, Santorini, Greece, July 14 2005.

[10] R. Ogier, F. Templin, and M. Lewis, "Topology Dissemination Based on Reverse-Path Forwarding (TBRPF)," RFC 3684, February 2004. [Online]. Available: http://www.ietf.org/rfc/rfc3684.txt

[11] C. Perkins, "IP Mobility Support for IPv4," RFC 3344, August 2002. [Online]. Available: http://www.ietf.org/rfc/rfc3344.txt

[12] U. Jönsson, F. Alriksson, T. Larsson, P. Johansson, and G. Maguire Jr., "MIPMANET - Mobile IP for Mobile Ad Hoc Networks," in *Proc. of MobiHoc 2000*, Boston, MA, USA, August, 11 2000, pp. 75–85.

[13] P. Srisuresh and M. Holdrege, "IP Network Address Translator (NAT) Terminology and Considerations," RFC 2663, August 1999. [Online]. Available: http: //www.ietf.org/rfc/rfc2663.txt

[14] P. Engelstad and G. Egeland, "NAT-based Internet Connectivity for On Demand MANETs," in *Proc. of WONS 2004*, Madonna di Campiglio, Italy, January, 18–23 2004, pp. 4050–4056.

[15] R. Droms, "Dynamic Host Configuration Protocol," RFC 2131, March 1997. [Online]. Available: http://www.ietf.org/rfc/rfc2131.txt

[16] N. Vaidya, "Weak Duplicate Address Detection in Mobile Ad Hoc Networks," in *Proc. of ACM MobiHoc 2002*, Lausanne, Switzerland, June, 9–11 2002, pp. 206–216.

[17] S. Nesargi and R. Prakash, "MANETconf: Configuration of Hosts in a Mobile Ad Hoc NEtwork," in *Proc. of INFOCOM 2002*, vol. 2, New York, NY, June, 23–27 2002, pp. 1059–1068.

[18] K. Weniger and M. Zitterbart, "Address Autoconfiguration on Mobile Ad Hoc Networks: Current Approaches and Future Directions," *IEEE Network*, vol. 18, no. 4, pp. 6–11, July/August 2004.

[19] E. Belding-Royer, *Routing Approaches in Mobile Ad Hoc Networks*. New York, NY: IEEE Press and John Wiley & Sons, 2003.

[20] S. Carl-Mitchell and J. Quarterman, "Using ARP to Implement Transparent Subnet Gateways," RFC 1027, October 1987. [Online]. Available: http: //www.ietf.org/rfc/rfc1027.txt

[21] C. Tschuding, R. Gold, O. Rensfelt, and O. Wibling, "LUNAR: a Lightweight Underlay Network Ad-hoc Routing Protocol and Implementation," in *Proc. of NEW2AN'04*, St. Petersburg, Russia, February, 2–6 2004.

[22] A. Tønnesen. (2004, December) Implementation of the OLSR specification (OLSR_UniK). Version 0.4.8. University of Oslo. [Online]. Available: http: //www.olsr.org/

[23] E. Borgia, "Experimental evaluation of ad hoc routing protocols," in *Proc. of IEEE PerCom 2005 Workshops*, Kauai Island, Hawaii, March, 8–12 2005, pp. 232–236.

[24] C. Gomez, D. Garcia, and J. Paradells, "Improving Performance of a Real Ad-hoc Network by Tuning OLSR Parameters," in *Proc. of IEEE ISCC 2005*, Cartagena, Spain, June, 27–30 2005, pp. 16–21.

In: Computer Networks, Policies and Applications
Editors: C. M. Berger and A. D. Streusse
ISBN: 978-1-61209-088-7

Chapter 4

THE IMPACT OF LICENSE AGREEMENTS ON ACCESS TO INFORMATION: A CHALLENGE FOR ACADEMIC LIBRARIES

Stacey L. Bowers
University of Denver, Sturm College of Law; Westminster Law Library,
Denver, Colorado, USA

ABSTRACT

Today's academic libraries are forced to enter into a significant number of licensing agreements every year for access to electronic resources. Vendors, or licensors, have the power to dictate the terms and conditions set forth in the license agreement. While libraries may attempt to negotiate and change certain terms, their attempts are often futile or met with strong resistance. Ultimately, the terms of these license agreements dictate the ability of libraries to provide access to electronic resources for their patrons. The license terms may limit or curtail access by certain patron groups, including alumni, walk-in users, the public, and other higher education institutions. The license agreement may also eliminate or restrict libraries' abilities to rely on the fair use provisions or engage in interlibrary loan under copyright law. Additionally, libraries may choose to or be forced for economic reasons to discontinue their print collection of titles that are also available electronically. These libraries may be left scrambling to provide access to these same resources if the licensor eliminates or changes its titles, years of coverage, or ceases to exist as an entity. As a result of these scenarios, libraries may find themselves unable to provide the level of intended access to electronic resources in accordance with their goals or mission statement. This inability to provide access can be directly tied back to the license agreement and its terms and conditions.

INTRODUCTION

Access is the right of a patron to enter a library and use its collections and resources openly and freely, including the right to use computer systems and online resources (Reitz, 2004). A core value of the American Library Association (ALA) is the concept of access to information. The ALA states, "All information resources that are provided directly or indirectly by the library, regardless of technology, format, or methods of delivery, should be readily, equally, and equitably accessible to all library users" (American Library Association Core Values of Librarianship, 2007). The ALA believes libraries should consider their goals, missions, and objectives, as well as the needs of their communities in determining the best way to provide access to electronic information (American Library Association Access to Electronic Information, Services, and Networks: An Interpretation of the Library Bill of Rights, 2007). Access to electronic information is a critical component of providing and making available resources to a library's patrons. Without access to electronic information, a patron is limited and hindered in his or her ability to engage in effective information retrieval.

An overriding commitment of academic libraries is to provide patrons with access to resources, whether print, electronic, or other formats. These patrons represent a diverse group of individuals, including the primary constituency generally comprised of the institution's students, faculty, and staff. Libraries may also view alumni and other higher education institutions as patrons. In addition, academic libraries at public institutions often serve the general public, a group that consists of a broad array of people ranging from an independent academic researcher to a local community member.

The mission statements of many academic libraries at both private and public institutions state that one of their goals is to provide patrons with access to information resources and services. While mission statements vary, the following represent a mix of various libraries' missions. A private academic law library states that its mission is to provide its students, faculty, and alumni, as well as members of the legal bar community and the public, "... with access to the broadest array of legal information sources" (Westminster Law Library Mission Statement, 2007). A public university's law library states, "The primary mission...is to evaluate, select, organize and provide access to information resources and services that support the instructional, research, and service programs of the students, faculty, and staff..." (University of Colorado Law Library Collection Development Policy, 2007). In addition to serving the needs of the greater university community, this law library also serves other groups, including the state's legal community, other academic institutions, and the general public (University of Colorado Law Library).

An academic library located on a campus serving three diverse public institutions, "...aspires to provide the best of both traditional and electronic information resources and services" (Mission of the Auraria Library, 2007). In addition to serving the patrons of the three institutions on the campus, the library also serves any resident of the state (Mission of the Auraria Library). A public university's academic library has a goal of sharing its resources with other higher education institutions, both nationally and internationally (University of Colorado at Boulder University Libraries Mission Statement, 2007). A private university library states that, "the library links information and the university community regardless of the location of the information or the individual seeking it" (University of Denver Penrose Library Frequently Asked Questions, 2007). Another private university library sets as one of its goals the provision of access to materials that are located both within and outside the library for its university community (Regis University Libraries Mission Statement, 2007). In addition, this library engages in community outreach by offering access to government publications and local area materials to the general public (Regis University Libraries Mission Statement).

What is readily apparent from this mix of mission statements is that academic libraries, whether serving a private or public institution, establish broad goals regarding access to information. While the primary goal remains serving the institution's main constituency of students, faculty, and staff, these libraries also believe in providing access to alumni, other academic institutions, and the public. In addition, academic libraries often view it as crucial to provide access both on-site at the library, and remotely regardless of where the patron is currently located.

While libraries can provide wide-ranging access to print materials, whether to their primary patrons, alumni, other institutions, or the public, these same libraries may not be able to meet this goal as it pertains to electronic materials. The ability to provide a broad range of access to electronic resources is often curtailed by outside factors. These hindrances include specific language in the license agreement limiting or denying the use of these electronic resources by certain patrons and/or eliminating the library's right to provide information from these resources through interlibrary loan, and economic constraints due to the cost of licensing fees. In addition, as libraries spend a larger percentage of their budgets on electronic resources, they may no longer collect those same resources in print format. As a result, the goal of providing broad-based access is curtailed.

Shift from Paper to Electronic Resources

With the advent and growth of materials that are available electronically, there has been a shift by libraries toward providing more and more access to resources only in electronic format. In particular, many libraries are cancelling subscriptions to their print journal collections when they are able to make the journals available to patrons through an electronic database option. This shift from paper to both paper and electronic, or to electronic only has occurred for a variety of reasons.

Many libraries have determined they cannot afford to maintain both the print and electronic versions of the same resource. As a result, libraries often determine which version to provide for patrons, and more frequently than not the electronic version is selected. For some libraries, it is a clear cost advantage to shift from paper to electronic resources. Many libraries license the content from electronic database vendors as a consortium. As a result of these consolidated purchases, libraries may be able to provide electronic access for less than the cost of print access (Roether and Koepf, 2005).

In addition to economic considerations, libraries may take into account space issues when making a determination as to whether or not to maintain print resources that are also available electronically. Available space and shelves are often at a premium in academic libraries. One way to solve the problem of limited shelf space is to move print copies of journals that are available electronically to off-site storage and also to discontinue the collection of print journals available through the library's electronic databases (Wallenius, 2007).

Another consideration that libraries may take into account when determining whether or not to retain the print version of a resource that is also available electronically is the level of recorded use (Roether and Koepf, 2005). Libraries may determine the extra cost and usage of space to maintain print versions of those journals most frequently used by students and faculty in their research is a prudent investment, even when more costly to subscribe to the print version rather than the electronic version (Roether and Koepf, 2005).

The shift from print to electronic resources brings both advantages and disadvantages. By moving to electronic resources, libraries can often provide a more in-depth collection of journal titles for patrons than they could in a print only format (Wallenius, 2007). Electronic resources generally allow libraries to provide broader access to its authorized users both on-site and off-site because remote access is available around the clock from any location. Another advantage of access through electronic resources is the constant availability to multiple patrons at the same time (Wallenius). In addition, electronic materials do not inhabit physical space in the library, which may allow libraries to add additional print titles to the collection that they did not previously have the space to accommodate. By shifting to more electronic resources, libraries can also meet the demands of many of their patrons for instant access, anytime and from anyplace.

Despite these advantages, there are also a number of disadvantages in the shift from paper to electronic materials. One disadvantage is the cost of licensing access to electronic resources. For many libraries providing electronic databases to its patrons may be cost prohibitive. Another significant disadvantage is that libraries have little or no control over the database because they do not own the content; libraries only license or lease access to these resources. The vendor chooses which content to keep, add, or delete from its database. Many licensing agreements allow the vendor to change the titles and years of coverage contained in its database offerings at its sole discretion. As a result of the vendor's ability to change its offerings on a whim, a journal title that is available during one subscription period may not be available during the next period. In the event that libraries have cancelled a print title in favor of electronic only access, they may find themselves with no copies of that particular title if the database vendor chooses to remove it from their list of available offerings. As a result, access to that particular resource is completely eliminated, both electronically and in print.

Another problem with electronic databases is the potential delay in publication of the content. There can be a significant time delay from publication of the print version to availability of the electronic version, depending upon the particular database vendor and title. In some instances, the electronic version may not be available for up to six months after publication of the print resource (Roether & Koepf, 2007). In those cases where the library only provides access through the electronic version, patrons are forced to wait for access to current information or must find an alternative means of access.

Relevant Copyright Provisions

In order to fully understand its rights and how best to negotiate the license agreement, the library must have a clear understanding of various copyright law provisions and how these provisions influence its ability to provide access for patrons. In particular, the library must understand the impact of licensing arrangements in regard to Section 109 of copyright law, the first sale doctrine. The library must also pay close attention to those license agreement provisions which impact its ability to rely on Sections 107 and 108 of copyright law, fair use and interlibrary loan. While the original intent of copyright law was to "balance the access rights of the public with the profit and control interests of copyright holders," license agreements have shifted that balance (Bartow, 2001, p. 833). It is now the copyright holder who maintains the power and in regard to electronic resources, that party is the database vendor. In today's reality, the license agreement terms override existing copyright law and the library is bound by the provisions of the license in regard to what it can and cannot do with an electronic resource (Johnson, 2004).

The library's rights under the doctrine of first sale are completely eliminated by the license agreement structure. Under copyright law this doctrine permits a library to purchase material and subsequently loan, resell, or gift that material to patrons (U.S. Copyright Act §109). However, when a library enters into a license agreement, it is not purchasing the electronic content; it is only licensing or leasing that material. As a result, the library has not engaged in an actual and outright purchase, and so the rights provided under the first sale doctrine do not apply to the electronic resource. In essence, this means the library cannot loan the material, resell the material, or give the material away as it can with print materials that it purchases outright. Thus, the license agreement structure curtails the library's rights under copyright law and its ability to provide broad access to the electronic resource. The consequence of losing the right of first sale under license agreements is that the library must comply with the license agreement terms, which means strictly controlling when and where access can be made available to patrons. The library cannot loan, resell, or gift any of the electronic content unless permitted under the license agreement, which is highly unlikely.

The fair use provisions of copyright law are essential to the library and its ability to provide access to patrons. Fair use allows the library, as well as its patrons, to make use of copyrighted material for purposes such as criticism, comment, news reporting, teaching, scholarship, or research (U.S. Copyright Act §107). Pursuant to the fair use doctrine, a library or its patrons can make copies of copyrighted material provided it falls within fair use. In making a determination of whether or not the library is engaging in fair use, the library must consider the following four factors: (i) the purpose and character of the use; (ii) the nature of the work; (iii) the amount used in proportion to the whole; and, (iv) the effect of the use on the potential market and value of the work (U.S. Copyright Act §107). As a result of the fair use provision, a library is often able to make photocopies of copyrighted works for their patrons when the patrons are using such materials for teaching, scholarship, or research. Without fair use rights, a library would be greatly hindered in its ability to serve and assist students and faculty in their studies and research. For example, if the license agreement contains a provision which eliminates the right of fair use, a faculty member would not be permitted to make multiple copies of the electronic content for his or her students for teaching purposes. A patron would be unable to make a copy of information contained in the electronic database for research or scholarship purposes. While some license agreements specifically eliminate fair use rights, it is even more typical for a license agreement to restrict or eliminate the library's right to engage in interlibrary loan transactions.

Section 108 of copyright law allows libraries to engage in the practice of interlibrary loan. This provision allows a library to make a copy of a work and distribute that copy to a user who has requested the work either from that library or through another library (U.S. Copyright Act §108(d)). The copy must become the property of the user who requested it, and the library must believe that the user will use the copy only for the purposes of private study, scholarship, or research (U.S. Copyright Act §108(d)). Furthermore, the library must prominently display appropriate copyright warning language where interlibrary loan orders are accepted and on the related order forms (U.S. Copyright Act §108(d)).

If the library enters into a license agreement which preempts its rights under the interlibrary loan provision of copyright law, the library will have a difficult time providing access to some of its patrons. The library will be prevented from printing a copy of any electronic content contained in that database and loaning it pursuant to Section 108. While the library may have retained its fair use rights which would allow for a copy to be made in certain situations, the lack of rights to engage in interlibrary loan preempts the ability of anyone who is not an authorized user under the license agreement to access the electronic content. As a result, "the tradition of libraries freely lending, borrowing, and exchanging portions of their collections with other libraries" is restricted or eliminated (Bartow, 2001, p. 826). When the library is not able to engage in interlibrary loan transactions as a result of the license agreement, many of its potential patrons endure a great disservice - severely limited access. This license restriction results in the curtailment of one of the foundations of libraries, interlibrary loan.

When a library enters into any license agreement, the first right that is stripped is its ability to rely on the doctrine of first sale. In addition, many license agreements also restrict or eliminate the library's ability to rely on Sections 107 and 108 of the copyright law. As a result, the library gives up control to provide the services needed and often demanded by its patrons. This loss can deter the library from its stated mission of providing access to information. Once the library has entered into the license agreement, it cannot rely on copyright law because it provides broader rights. The library must abide by the language of the license agreement, no matter how severely such language restricts the library in its ability to engage in fair use and interlibrary loan transactions, and ultimately provide the access it has promised.

License Agreements

A direct consequence of this continuing shift from paper to electronic resources is the licensing agreement. The license agreement is simply the contract entered into between the licensee, (i.e., the library), and the licensor, the vendor providing the electronic database and access to its content. Electronic databases are comprehensive collections of digitized information generally related to a specific area or field of study and are designed for ease of access (Reitz, 2004). The database vendor typically leases or purchases the electronic content and makes it available using their specific search software or interface (Reitz). Vendors protect their rights to the electronic product by requiring libraries to enter into the license agreement in order to gain access to the content. In most cases, the vendor drafts and provides the license agreement to the licensee. In some instances the licensee may be able to negotiate various provisions of the agreement, and in other cases the license may not be subject to negotiation.

Like most legally binding contracts, the license agreement contains numerous terms and provisions that govern the contractual relationship between the parties. A number of these provisions impact a library's ability to provide its patrons with access to this electronic information. The specific provisions that are most likely to impact who the library can provide access to and how such access can be provided include: (i) the definition of authorized user; (ii) the definition of authorized site; (iii) the scope of permitted uses; and,

(iv) whether or not the library has kept its fair use and interlibrary loan rights pursuant to copyright law. Additionally, the provision addressing whether or not the library has the right to create an archival data file may determine the library's ability to provide on-going access to the electronic information.

The definition of authorized user is critical in every licensing situation and determines who is legally permitted to use the electronic database. The library must have a clear understanding of its patrons and should seek to define the term authorized users based on this knowledge. The library should also take into account its stated goals and mission when determining who should be included as authorized users. In most license agreements, the term authorized user includes students, faculty, and staff affiliated with the academic institution, and walk-in patrons as permitted by the library. In some situations, the license agreement may attempt to limit authorized users to students, faculty, and staff affiliated with a particular unit of the institution, such as a law school or business school. In many instances, this standard definition of authorized user does not meet the needs, goals, or mission of the library, or the expectations of its patrons. The library may be committed to providing broader access and want to include visiting faculty, adjunct faculty, alumni, the public, or other groups, but may be unable to do so as a result of the limited definition of authorized user pursuant to the license agreement.

The inclusion of walk-in patrons in the definition of authorized users generally allows the library to offer access to visiting or adjunct faculty, alumni, or the public. However, if the license agreement stipulates walk-in users at the library versus at the institution as a whole, the library would only be permitted to offer access to those persons actually present in the library and using a computer in the facility. As a result, the license prevents the visiting or adjunct faculty from accessing the resource from their office computer or from anywhere other than the library.

More often than not, alumni are not specifically listed in the definition of authorized user. This means access can only be provided to this group of patrons in a limited manner, if at all. If the license agreement allows for walk-in users, then alumni who live locally could walk into the library and utilize the electronic resource on-site. However, many alumni do not remain in the area after graduation and although the library may have a mission of providing access to them, it will not be able to meet that goal because they cannot utilize the electronic resource from any location other than on-site. In those limited instances where the license agreement does specifically allow for access to alumni, it may still be difficult to make such access available to that user group remotely. It is often cost prohibitive to even consider including alumni in the group of authorized users. The vendor may charge a premium for allowing this type of broad-based patron access to persons who are not currently affiliated with the institution.

Unlike the print resource, which any library patron who is permitted to use the library may access and utilize, the electronic resource is restricted to only those authorized users as defined in the license agreement (Roether and Koepf, 2005). As a result, the library should attempt to define authorized user as broadly as possible so that it can meet the overriding goal of providing access to all its patrons. However, the definition of authorized user often conflicts with or limits the library's stated goals or mission. As a result, resources that were once readily accessible to any patron of the library now reside behind the wall of a proprietary database that is accessible by far fewer patrons (Wand, 2005). With access provided electronically in lieu of print, these resources may be forever lost to those patrons who do not fall within the authorized user definition.

Another concern for the library is the omission from the definition of authorized users of a group of particular patrons, such as staff, visiting faculty, or walk-in users. In that case, the library may find itself in the precarious situation of not being able to offer the electronic resource to them or having to attempt to renegotiate that portion of the license agreement. The library should never assume that unlisted patron groups are covered in the definition of authorized users. No matter the end result, the library must abide by the definition of authorized user and ultimately, it is this definition that will determine who can and cannot access the electronic information.

In conjunction with authorized user is the concept of authorized site. This license agreement provision determines the location from which users may access the electronic database — on-site only, or on-site and remotely. If the library desires to make the electronic content available to its authorized users remotely, the license must specifically state that remote access is permitted (Harris, 2002). While most license agreements allow the library's primary patrons, students, faculty, and staff, to access the electronic resource off-site, the license does not often allow other library users, alumni and the public, to do so. As a result, alumni and public patrons will be relegated to using the electronic database on-site at the library. While the library may state that it provides access to alumni and the public, these site restrictions will limit access to only those persons who have the ability to visit the library. While many license agreements allow on-site and remote access for the library's primary patrons, this is not always the case. In some instances, the library will have to negotiate the right to provide off-site access to the electronic database even for students, faculty, and staff. This may result in additional licensing fees. Due to these added costs, the library may be reluctant or unable to provide any off-site access, even to its primary patrons.

Libraries must also pay careful attention to the license agreement and the permitted uses it stipulates for the electronic resource. In some instances, the database vendor, through the license agreement, will limit use of the electronic resource to academic research or educational use only (Lowry, 1993). In that case, patrons of the library may only utilize the electronic resource when they are engaged in academically based research or educational purposes. This means that even the library's primary patrons cannot use the resource for work-related research, an internship experience, or for pleasure. Additionally, if the license only allows for academic research usage or educational use, alumni, walk-in users, and the public would be prohibited from utilizing that resource unless engaged in some type of academic or educational research. As a result of this type of license agreement provision, the library is required to police patrons who have access to the electronic information and the manner in which they use the resource (Bartow, 2001). Additionally, limitations on permitted uses impact the librarians and their ability to utilize the electronic resource to assist patrons. This will require the librarians to determine if the patron is engaged in a permitted use and to determine whether or not they can utilize that particular electronic resource to assist the patron. License agreement provisions that limit permitted uses of the electronic resource are very difficult to govern. They require more stringent standards regarding access to ensure that those persons engaged in use of the resource are allowed to do so. Otherwise, the library may find itself in violation of the license agreement.

Another concern with any license agreement is its structure as a license (or lease) and not an outright purchase. By characterizing the transaction as a license, the database vendor is preventing the library from engaging in a purchase; therefore, the library cannot simply do what it desires in compliance with copyright law with the electronic resource in the ways that are possible with a print resource. As a result, license agreements compromise library patrons' access to information in a way that is not possible with books or paper journals (Bartow, 2001).

In addition to eliminating first sale rights, many license agreements restrict or override a library's ability to engage in fair use of the electronic content or engage in interlibrary loans of such content. While these rights are granted under copyright law, the license agreement can contract away those rights from the library. Some license agreements restrict or eliminate entirely the library's right to engage in interlibrary loan transactions. The license may stipulate that the library can only engage in interlibrary loan within the institution or only if the library also owns a print version of the electronic material. In other cases, the license may explicitly prohibit all interlibrary loan transactions no matter the reason — academic use or otherwise. The library must carefully review any license agreement provision that affects copyright law and attempts to restrict or eliminate those rights. The library must then make a conscious determination as to whether or not those license agreement infringements are acceptable to the library and its goals and mission.

A final concern is whether or not the library has been granted the right to archive an electronic copy of the database and its content. Unlike previous electronic products, such as CD-ROMs and disks, when the electronic content is stored and only available through remote Internet access, the library no longer holds or owns a tangible product (Klinefelter, 2001). Since the library has no control over the electronic resource, it has no ability to preserve that information (Klinefelter). While some licenses permit the library to make an archival copy of the electronic content, in many cases the database vendor retains all archival rights (Klinefelter). If the library is not permitted to store an archival copy of the electronic content, it may find itself without access to any of that information, particularly in the event where the vendor chooses to delete specific titles or years of coverage from its database collection or ceases to exist as a corporate entity. As a result, the library can no longer provide access to any of its patrons, even if they meet the definition of authorized user.

Impacts and Consequences

What is abundantly clear is that while more information continues to become available electronically, license agreements impede a library's ability to provide access to that same electronic information. As is evident from the examination of various libraries' mission statements, a main goal of most academic libraries is to provide broad access to resources, whether print or electronic, for all its patrons. While the library may have set forth a mission of providing broad-based access to all resources for its patrons, it will often not be able to meet those goals due to the various licensing agreements into which it has entered.

These license agreements impede a library's ability to provide access in a number of ways. One main way licenses curtail access is through the definition of authorized user. Alumni, a group generally not specifically listed in the definition of authorized user, are an important group in supporting the library post-graduation (Horava, 2007). As a result, many libraries commit to providing alumni access to their resources. Some libraries continue to allow full or limited borrowing from the print collection, reference support, and access to electronic resources (Horava). However, while libraries are promising broad access to the library's resources for alumni, in many instances the library is unable to provide access to its electronic resources for this group unless they are willing to utilize the resource on-site at the library. This may be impossible for alumni who have relocated or not ideal for alumni who prefer to access information from wherever they are at any given moment. Additionally, alumni are often not familiar with the implications of license agreements and the ways in which they limit access to various user groups. Due to these impediments to access, alumni may become frustrated and less willing to support the library after graduation if they believe promises cannot be honored.

Another way in which license agreements impact the library's ability to provide access to patrons is by stipulating that authorized users can only access the electronic database while on-site at the library or the greater institution. As a result of this type of restriction, authorized users cannot access information at their convenience from any location. This type of site restriction significantly affects users who expect electronic information to be available to them at all times no matter where they are. It also has the potential to impact institutions and they way in which they provide access to their distance students. If a license agreement does not allow for off-site access, distance students may be denied the ability to access materials needed for their educational experience.

License agreements also impinge on a library's ability to provide access by restricting or eliminating various copyright privileges that a library would normally be entitled to rely on. License agreements often eliminate the library's right to engage in interlibrary loan transactions. One primary reason that database vendors eliminate a library's ability to engage in interlibrary loan transactions is for economic reasons. If the database vendor allows libraries to engage in interlibrary loan, the vendor may lose substantial dollars because a third party who might choose to enter into a license with the vendor no longer needs to do so since they can simply request the same material via interlibrary loan. By eliminating this ability, the library will no longer be able to serve outside patrons. Thus, the library cannot provide access to other institutions and their patrons, and once again access has been severely impeded by the license agreement.

A related problem as a result of these proprietary electronic databases is their ability to limit access to information to those persons who are independent researchers or scholars. Since these persons are considered non-affiliates of the university they are relegated to only obtaining access to information as a walk-in patron, if that is permitted at the institution, or possibly by paying out of pocket for such access. As is the case with many private institutions, the only people permitted on-site access to the library and its resources are current students, faculty, and staff. On occasion, these institutions may permit continued access by alumni, but that is not always the case. Additionally, these institutions often do not permit access to the library for public patrons and in those circumstances when they do, the public patron must pay a daily use fee to work on-site at the library. In the past when the majority of content was available in print form, these patrons could request the material through interlibrary loan and access it in that manner without suffering monetary penalties. However, with the advent of strict licensing agreements, libraries are often not permitted to engage in interlibrary loan of electronic content and resources. As a result, a whole segment of the population, independent researchers and scholars, are denied access to pertinent information (Wand, 2005).

Due to these license agreement restrictions, libraries and their librarians may look for creative ways to continue to provide access to all their patrons. In some cases, librarians attempt to work around the terms and limitations of the license agreement by performing a search in the electronic database for a patron. Due to the fact that librarians are permitted access to that electronic database, they believe that engaging in a search for a patron who does not fall within the definition of authorized user is acceptable. However, by engaging in this type of activity the library may very well be in breach of its license agreement.

What is apparent is that no matter the library's goals and mission, it is unlikely that it will truly be able to live up to those statements as a result of the limitations imposed by license agreements. While some licenses may impose greater restrictions or limitations than others, most license agreements will impede a library's ability to provide access to its patrons in one form or another. While these impediments to access may seem small when looked at individually, when examined in the aggregate, they are large and significant. In a world where more and more information can only be accessed electronically, if license agreements continue to impede libraries' abilities to provide access to all its patrons, the divide between those people who have access to electronic databases and their content and those people who do not will only continue to grow.

CONCLUSION

A core value of librarianship, equal access for all users, is challenged by this shift to electronic content and the resulting license agreement structure. It is essential that libraries pay closer attention to this issue of access to electronic information. Libraries should be greatly concerned with continuing to provide the same level of access to their patrons as was available before the advent of and continual shift to electronic resources. Unfortunately, the nature of licensing agreements is preventing libraries from doing just that. In this era of Google, instant messaging, and instant access, many library patrons want access to everything from wherever they are located at that given time (Horava, 2007). However, this shift to electronic delivery of information actually impedes access for many patrons of the library. Database vendors have the ability to restrict and control access to their information as a result of the same electronic and instant access technology on which many library patrons have come to rely (Bartow, 2001).

Libraries must engage in close scrutiny of every license agreement into which they enter. They must determine how the various provisions of the license will impinge on their ability to provide access to patrons, whether their primary constituency or those that fall outside of the group, such as other higher education institutions, alumni, and the public. In particular, attention must be paid to the provisions that impact the library's right to engage in fair use of copyrighted materials and its ability to engage in interlibrary loan transactions. These rights have become implicit when considering print materials, yet are often amorphous when dealing with electronic resources. Access to electronic information will continue to be more limited as we move into the future unless libraries begin making changes to the terms of the license agreements that they enter into for electronic content.

In order to provide broad-based access for all its patrons, libraries must continue to fight for these rights. The first step in this fight is to negotiate license agreement terms that allow the library to provide broader access or to negotiate for the purchase of the electronic content as opposed to the licensing or leasing of such content. It will take a coalition of libraries acting as a unified front to begin this process of change. Currently, database vendors hold the upper hand and have the ability to force their terms of access upon libraries. Were libraries to present a more unified front and require database vendors to provide access on their terms, a shift may begin to take place. Then the libraries may find themselves the more powerful party in the licensing arrangement.

An example of libraries unifying is the "Shared E-Resource Understanding" (SERU) project. Recently, a coalition of 21 librarians and 8 publishers agreed to use the SERU document, which sets forth a list of points that each side agrees to abide by (Glenn, 2007). The parties participating in this pilot initiative agree to deal in good faith if or when a dispute arises between them and to abide by a list of stipulated guidelines or understandings with the only legally binding contract being the purchase order (Glenn). This project is still in its infancy and the publishers participating include an open access journal publisher and university presses. It remains to be seen whether this idea will work and if it could be transitioned into and implemented by large database vendors in licensing agreement situations.

"Today, when the Web gives society the illusion that all information is available to everyone, more and more of the latest and most highly vetted research findings are locked up in expensive, proprietary databases accessible by fewer and fewer individuals" (Wand, 2005, p. 30). We, as a society, must have access to this electronic content in order to contribute to new knowledge and thinking (Wand). Libraries must persevere in their efforts to provide access to electronic resources for all their patrons. If not, access will continue to become more and more limited and huge segments of the population will be unable to access much needed resources and knowledge. Libraries cannot rely on others, particularly database vendors, to provide current and on-going access to the information locked away in electronic databases. The party who cares about access and is willing to provide access to all users is the library. In order to meet their goals and missions, libraries must step to the forefront in the fight against the contemptible limitations that database vendors impose with their license agreements.

REFERENCES

American Library Association Access to Electronic Information, Services, and Networks: An Interpretation of the Library Bill of Rights. Retrieved October 2, 2007, from http://www.ala.org/Template.cfm? Section=interpretations and Template=/ContentManagement/ContentDisplay.cfm and ContentID=133993

American Library Association Core Values of Librarianship. Retrieved October 2, 2007, from http://www.ala.org/ala/oif/statementspols /corevaluesstatement/corevalues.htm

Bartow, A. (2001). Libraries in a digital and aggressively copyrighted world: Retaining patron access through changing technologies. *Ohio State Law Journal, 62,* 821-834.

Glenn, D. (2007, September 21). Librarians and publishers try out a plan to simplify negotiations over electronic resources. *The Chronicle of Higher Education.* Retrieved September 21, 2007, from http://0-chronicle.com.bianca.penlib.du.edu/daily/2007/09/2007092103n.htm

Harris, L. E. (2002). *Licensing digital content: A practical guide for librarians*. Chicago: American Library Association.

Horava, T. (2007). Licensing e-resources for alumni: Reflections from a pilot project. *College and Research Libraries News, July/August,* 437-441.

Johnson, P. (2004). *Fundamentals of collection development and management*. Chicago: American Library Association.

Klinefelter, A. (2001). Copyright and electronic library resources: An overview of how the law is affecting traditional library services. *Legal Reference Services Quarterly, 19,* 175-193.

Lowry, A. (1993). Landlords and tenants: Who owns information, who pays for it, and how? *Serials Librarian, 23(3-4)*, 61-71.

Mission of the Auraria Library. Retrieved September 26, 2007, from http://library/ auraria.edu/ aboutus/whoweare/mission.html

Regis University Libraries Mission Statement. Retrieved September 26, 2007, from http://www.regis.edu/library.asp?page=about.overview.mission

Reitz, J. M. (Ed.). (2004). *Dictionary for library and information science*. Westport, CT: Libraries Unlimited.

Roether, D., and Koepf, L. (2005). Information may become freely available, but information is not free, *Best Practices, Winter,* 1-5.

United States Copyright Act, 17 U.S.C.A. §107 (ThomsonWest 2005).

United States Copyright Act, 17 U.S.C.A. §108 (ThomsonWest 2005).

United States Copyright Act, 17 U.S.C.A. §109 (ThomsonWest 2005).

University of Colorado at Boulder University Libraries Mission Statement. Retrieved August 12, 2007, from http://uclibraries.colorado. edu/dea/mission.htm

University of Colorado Law Library Collection Development Policy. Retrieved September 26, 2007, from http://www.colorado.edu/Law/ lawlib/ts/index_print.htm

University of Denver Penrose Library Frequently Asked Questions. Retrieved August 12, 2007, from http://penrose.du.edu/FindIt/HelpWithResearch/ FAQ/index.cfm?id1=1 @id3=47

Wallenius, L. I. T. (2007). Are electronic serials helping or hindering academic libraries? *Acquisitions Librarian, 19,* 75-82.

Wand, P. (2005). On my mind: Inaccessible information: A strategic solution. *American Libraries, 36(5),* 30.

Westminster Law Library Mission Statement. Retrieved August 12, 2007, from http://www.law.du.edu/library/content.cfm?pg=mission

In: Computer Networks, Policies and Applications
Editors: C. M. Berger and A. D. Streusse
ISBN: 978-1-61209-088-7

Chapter 5

BUNDLING RESIDENTIAL TELEPHONE, INTERNET, AND VIDEO SERVICES: ISSUES FOR CONGRESS*

Charles B . Goldfarb

ABSTRACT

Technological advances and deregulatory actions now allow consumers to obtain their local and long distance telephone services, their high-speed Internet services, and their video services from competing technologies. The convergence of previously distinct markets has required companies to seek strategies for holding on to their traditional customers while seeking new ones. One of those strategies is for companies to offer bundles of "traditional" and "new" services at a single price that often represents a discount off the sum of the prices of the individual services. These bundled service offerings are favored by many consumers. They provide the convenience of "one stop shopping" and in some situations, by providing the full panoply of services at a fixed price, make it easier for consumers to comparison shop. They also are favored by many providers because they tend to reduce "churn" - the rate at which customers shift to competitors - and allow providers to exploit economies of scope in marketing.

But bundling also can create public policy issues for Congress. The bundled offerings typically provide some combination of interstate telecommunications services, intrastate telecommunications services, and non-telecommunications services (information services, video services, and even customer premises equipment) for a single price. The federal Universal Service Fund - the federal subsidy program that assures affordable telephone rates for high-cost (rural) and low-income telephone customers as well as for schools, libraries, and rural health facilities - is supported by an assessment on interstate telecommunications revenues only. But it is difficult to identify the portion of revenues generated by a bundled service offering attributable to the interstate telecommunications portion of that bundle. There is no unambiguous way for providers to assign a portion of the bundled price to interstate telecommunications services or for fund administrators to audit that assignment. In addition, some taxes are assessed upon one or more, but not all, of the services included in various bundled service offerings. This creates the same assessment and auditing problem for these taxes

* This is an edited, reformatted and augmented edition of a United States Congressional Research Service publication, Report RL32232, dated February 17, 2004.

as exists for the federal Universal Service Fund. This has important policy implications at a time when many Members of Congress seek to shelter Internet services-which often are included in these bundles- from taxation without placing any group of providers at a competitive advantage or disadvantage.

Some observers have been concerned that bundled service offerings could have anticompetitive consequences if they foster industry consolidation or if a provider has market power for one of the services in its bundled offering and can use that offering to tie that service to a competitive service in a fashion that reduces competition for the competitive service.

Leaders in both the House and the Senate Commerce Committees have announced that in the 109" Congress they plan to review and reform the 1996 Telecommunications Act (P.L.104-104) in light of the market convergence that underlies the trend toward bundling.

Technological advances and deregulatory actions now allow consumers to obtain their local and long distance telephone services, their high-speed Internet services, and their video services from competing technologies. Companies that in the past sold a narrow suite of services in relative insulation from competition now are actively entering new service markets and also facing entry by others into their traditional markets. The convergence of previously distinct markets has required companies to seek strategies for holding on to their traditional customers while seeking new ones. One of those strategies is for companies to offer bundles of their "traditional" services and "new" services - typically at a single price that represents a discount off the sum of the prices of the individual services.

Today, most incumbent local exchange carriers ("ILECs"),[1] competitive local exchange carriers ("CLECs"),[2] wireless carriers, cable companies, and satellite television companies have bundled service offerings that compete, to varying degrees, with one another.

Leaders in both the House and the Senate Commerce Committees have announced that in the 109' Congress they plan to review and reform the 1996 Telecommunications Act (P.L. 104-104) in light of the market convergence that underlies the trend toward bundling.[3]

The Market Forces Driving Bundling

The trend toward bundled service offerings is driven by both demand and supply.

Bundled service offerings are favored by many consumers. They provide the convenience of "one stop shopping" and, to the extent that competitors' packages include the same (or a very similar) bundle of services, can make it easier for consumers to comparison shop, calculate their total telecommunications and media expenditures, and switch from one provider to another by switching just a single account.[4]

According to J.D. Power and Associates,[5] the share of households that report bundling at least their local and long distance services with one carrier has increased from 26% in 2002 to 40% in 2003, an overall increase of more than 10 million households, and customers who bundle services report higher overall satisfaction than those who are not bundling services. The Yankee Group made a similar finding - even though bundles have only been available in many parts of the country within the past year, a third of Americans already receive long distance and local service from the same company.[6] According to Wayne Huyard, president of MCI's mass markets division, half of MCI's consumer-side revenues comes from the 3.5 million customers of its bundled service and that proportion is expected to increase to 75% by 2005.[7]

But the bundling phenomenon extends far beyond the simple packaging of local and long distance telephone services, and is driven by providers as much as by consumers. Although bundled offerings can make comparison shopping easier, they also tend to foster customer loyalty, thereby reducing "churn" - the rate at which customers discontinue service (in order to shift to competitors). They also allow providers to exploit marketing economies of scope. As providers enter new markets, they can market both their traditional services and their new services with a consolidated sales and marketing force and campaign.

Until recently, most households had no alternative to their local exchange carrier for local telephone service and no alternative to their local cable system for subscription video service. The ILECs and cable systems enjoyed relatively stable relationships with their customers. Customer chum was extremely low because in most situations there were government-imposed prohibitions on other providers offering competitive service, leaving customers with no alternative providers to turn to. But today, CLECs, wireless carriers, and cable systems offer local telephone service in competition with the ILECs, and satellite systems offer subscription video service in competition with cable systems. TI, FCs and cable systems now face rising levels of customer chum,[8] analogous to what AT and T and its long distance competitors have experienced with their customers since MCI, Sprint, and a host of smaller facilities-based carriers and resellers entered the long distance market in the 1980s, and what the wireless carriers have begun to experience as well.[9] Competitive entry and market convergence have increased the chum rate for all providers.

The costs associated with customer chum are substantial . Providers face up-front costs to capture and serve a customer. To serve that customer profitably, the provider must recover these costs before the customer changes provider. These costs include the acquisition (marketing and sales) costs associated with gaining new customers, retaining existing customers, and winning back customers who leave for another provider. They also include costs associated with connecting the customer to the provider's network and activating service-providing the wireline or wireless connection, incorporating customer data in the provider's operating support systems, sometimes placing equipment on customer premises, etc. If the provider faces high churn rates, absent a substantial initial customer charge (such as a connection charge), it may fail to fully recover these up-front costs. One industry analyst has estimated that wireless providers have per customer acquisition costs of $ 150 to $300 and payback cycles as long as 14 months.[10] Another industry analyst estimates residential wireless per customer acquisition costs of $300 to $425.[11] But the wireless industry has annual customer churn rates in the vicinity of 30%, which are expected to grow with the recent implementation of local number portability.[12] Thus, analysts say wireless carriers currently are not able to fully recover their acquisition costs for a substantial portion of customers . The same is true for other telecom, Internet, and video service providers.[13] The higher the churn rate, the less likely providers will be able to fully recover their up-front costs from each customer - at the same time that pressure is placed on providers to increase their efforts (and costs) for acquiring and retaining customers.[14]

The high costs associated with churn have spawned an industry of market researchers to help firms identify customer purchasing patterns and construct strategies for minimizing chum.[15] According to these market researchers, increasing the number of services included in the bundle tends to reduce churn.[16] Bundled service offerings therefore tend to provide an advantageous strategy for large companies that are able to offer a broad array of services.

Verizon has announced that the company plans to offer consumers as many services as possible in its bundled offerings, with traditional voice options supplemented by wireless, video services, and high-speed Internet; according to Jill Wagner, Verizon vice president of consumer marketing, "It's not just the [local companies] and the long distance providers. You have to throw in the six wireless providers, and you have to throw in the cable companies. That's the market.[17] BellSouth has the same perspective. Lisa Fox, BellSouth's director of consumer marketing, has stated: "Because we can sell them local, long distance, data, wireless and - soon - video all on one bill, that's really proved to be a good retention tool for us. Customers can't find that in our region with anyone else today.[18]

As these large bundled service offerings have grown in popularity, companies with narrower capabilities that have traditionally offered stand-alone services have had to partner with larger companies in order to participate in a market environment that favors bundling.[19]

Since identical bundled offerings facilitate comparison shopping, providers have a strong incentive to differentiate or distinguish their bundled service offerings from their competitors' offerings. Otherwise, their primary way to hold on to customers is to keep cutting prices. The best way to differentiate a bundle is to include a service that competitors either cannot offer at all or cannot offer at the same quality, cost, or convenience.

WIRELINE-OF BUNDLING STRATEGIES

Most wireline providers - ILECs and CLECs - offer bundled local and long distance telephone services. Unadorned by special features, long distance service and (increasingly now) local service have become commodities, subject to fierce price competition and high churn levels.[20] Although many CLECs initially attempted to enter the residential market by using their own network facilities or by reselling the ILECs' retail products, in almost all cases CLECs have abandoned those approaches as not competitively viable. Instead, most CLECs provide residential local service by leasing network facilities from the ILECs, in particular by leasing the unbundled network element ("UNE") known as UNE-platform or UNE-P,[21] under terms set out in the 1996 Telecommunications Act, as implemented by the FCC and state public service commissions.[22] One group of CLECs has estimated that in the second quarter of 2003 80% of all residential local service offered by CLECs as part of bundled service offerings was provided using UNE-P leased from the Regional Bell Operating Companies ("RBOCs")[23]. Some ILECs have attempted to differentiate their bundled offerings from CLECs by including vertical features such as voice mail and privacy management (to block telemarketers) that are not part of the UNE-P they are required to provide to CLECs but that CLECs are unlikely to be able to offer at equal quality or cost on their own.[24]

Another way that ILECs attempt to differentiate their bundled service offerings from CLEC offerings is to offer high-speed Internet access service by using DSL technology to provide both voice and data services over the existing copper telephone lines. Although in many geographic locations it is not feasible for CLECs to deploy their own DSL equipment to serve residential customers [25] the DSL equipment is not available as a UNE, nor as part of the UNE-P. As a result, in many geographic areas, CLECs do not, themselves, offer bundled voice and DSL service. Even Covad, which specializes as a provider of high-speed Internet access and continues to expand its footprint, will have collocated DSL equipment in only 2,000 (out of more than 10,000) ILEC central offices by mid-2004, and will be able to reach somewhat less than half (under 50 million) of U.S. households[26]. Most residential customers seeking to receive both voice service and high-speed Internet access service over their telephone lines have the following choices: to receive both voice and Internet access service from their ILEC; to receive voice service from their ILEC and high-speed Internet service from a provider such as Covad, with that Internet service provider collocating its own DSL equipment in the ILEC's network and leasing the data portion of the ILEC line serving that customer at negotiated prices that need not reflect costs;[27] or to receive the two services through a "line splitting" arrangement [28] under which a CLEC that specializes in offering high-speed Internet access (data) service, and another CLEC that specializes in offering voice services, could jointly use the unbundled loop to provide the customer both voice and Internet access services. This last option has allowed some CLECs, including AT & T and MCI, to respond to the ILEC's bundled local/long distance/high-speed Internet access offerings by entering into contractual marketing relationships with Covad, to offer a similar bundled service. But it is more complex - and more expensive - to coordinate line splitting than for an ILEC to offer the two services using its own facilities.

According to a December 5, 2003 SBC press release,[29] Long distance and DSL help [SBC] reduce [its] chum:

- Adding long distance to an access line reduces the company's chum rate by 9 percent.
- Churn drops by 61 percent when a DSL line is added to an SBC bundle.
- Together, long distance and DSL reduce chum by 73 percent.

Wireline, Wireless, Video Bundling Strategies

Today, bundled offerings have expanded far beyond the telephone and high-speed Internet access services traditionally offered by ILECs and CLECs, to include video and wireless services as well. Initially, each provider's bundles tended to be limited to the combination of services that could readily be provided by the firm's underlying network technology, and thus each provider tended to offer different combinations of services. Traditional circuit-switched public telephone networks, coaxial cable television networks, wireless networks, Internet protocol networks, and satellite systems each have their advantages and limitations with respect to services offered. To minimize market fallout from the limitations of their chosen technologies, however, providers increasingly are teaming with companies that have different underlying network technologies in order to provide complete bundled offerings.

For example, the incumbent local telephone companies have been able to quickly enter the long distance market as soon as they received government approval to do so.[30] Verizon already serves 15.9 million long distance customers, SBC 11.5 million, and BellSouth 3.4 million.[31] But to date it has not proved viable for local telephone companies to use their circuit-switched telephone networks to offer video services. Instead, in order to compete with the bundled offerings of cable operators, many ILECs are entering into partnership arrangements with satellite companies to market satellite television services as part of a telephone company bundled offering. Both Verizon and BellSouth have entered into marketing agreements with DirecTV to begin offering subscription video services as part of their bundled service offerings in 2004 [32]. Similarly, SBC has entered into a co-branding deal with Echostar, also beginning in 2004, to offer "SBC Dish Network".[33]

Similarly, local cable television systems' fiber optic platforms have helped cable companies become the largest providers of high-speed Internet access (through the use of cable modems). As of September 30, 2003, 15 million cable customers received cable modem service.[34] But many cable systems have not yet made the network upgrades needed to offer telephone services ; as of September 30, 2003, there were 2.5 million residential cable telephony customers in the United States.[35] Recently, however, a number of cable systems have announced that rather than undertake the expensive investment needed to upgrade their coaxial cable networks to provide telephone service - and have to continue to pay high access charges to the 11 FCs to terminate calls - they plan to use Internet Protocol technology to offer voice services, in some cases jointly with long distance carriers.[36] Cable companies have undertaken this bundling strategy at least in part as a customer retention strategy against the satellite companies.[37] Similarly, some cable companies are deploying video-on-demand service to capture and maintain customers.[38]

According to Jeffrey Halpern, an analyst at Sanford C. Bernstein and Co.,[39] wireless is the key to distinguishing phone companies from their cable television rivals, many of which have phone and Internet access packages but don't have wireless offerings. The ILECs that have wireless joint ventures (SBC and BellSouth jointly own Cingular Wireless; Verizon has a joint venture with Vodafone Group that offers Verizon wireless service in the U.S.) have expanded their bundled service offerings to include wireless options that neither the cable companies nor the CLECs can so readily offer. There appears to be strong market pressure on these cable companies and CLECs to establish relationships with wireless carriers not affiliated with the ILECs to offer a bundled service that includes wireless service. AT & T is testing a bundled package that includes wireless service from its former wireless unit, AT & T Wireless Services.[40] But there are a limited - and apparently shrinking -number of unaffiliated wireless carriers. Indeed, the announcement on February 17, 2004 that AT & T Wireless has accepted an acquisition bid made by Cingular [41] could remove AT & T Wireless as an independent source of wireless service for AT & T if the acquisition is completed.[42]

Similarly, wireless carriers now are offering bundled packages that include local, long distance, and Internet access services, and satellite providers are offering various video packages and experimenting with non-video services. These companies seek bundle components that will reduce churn .

Table 1 presents a sample of bundled service offerings of major providers prepared by one industry observer. It is presented with the caveat that some of the specific bundles, prices, and geographic reaches listed are already out of date because virtually every week some provider is either expanding its bundle, extending the geographic area in which it is offering its bundle, or changing its prices.

THE PRICING OF BUNDLED SERVICE OFFERINGS

The pricing structure of bundled offerings tends to follow a few patterns. Bundled telephone service typically includes unlimited local, local toll, and long distance services at a single flat rate. It sometimes is difficult for consumers to compare that single rate to the sum of the rates of the components because the components (especially long distance and local toll service), when sold as standalone services, usually are sold on a usage (rather than flat rate) basis. Typically, the flat rate bundle is the cheaper option for consumers who are heavy telephone users, but the more expensive choice for consumers who are light users. According to a Wall Street Journal article,[43] "Some people who don't make a lot of calls and don't want services like call waiting soon discover that most unlimited packages, which are geared to high-end customers, aren't economical for them." This partially explains why MCI's Neighborhood product, "which costs $50 to $60 a month in most areas, loses about half its new customers within the first six months, though turnover drops after that. [44]

ILECs and CLECs that supplement their wireline telephone bundles with non-wireline telephone services, such as high-speed Internet access or wireless service or even video service, typically offer a bundle of their "traditional" services at a fixed price and then allow customers to add to that bundle by paying flat prices for additional services, with the prices for those additional services typically being lower when purchased as part of the bundle than the stand-alone prices for those additional services. Similarly, cable companies typically will supplement their subscription television offerings with high-speed Internet access and telephone services that have a separate add-on price that is lower than the stand-alone price for those services.

According to a *New York Times* article,[45] Cable companies, which face little competition from rival cable companies in many markets, have a great deal of leverage in pricing and are eager to expand their universe of high-speed Internet customers because the business has a higher margin than the video business. Cable customers who buy both the video package and high-speed Internet access pay somewhat less than customers who buy only Internet service." A perhaps more nuanced explanation for this pricing behavior is that cable companies' only competitors for subscription television are the satellite companies, which in most circumstances cannot offer their customers competitive high-speed Internet access service. Cable companies thus can reduce competitive churn by offering high-speed Internet service at a discount that is available only when the customer also takes cable service. For example, cable customers who switch to DirecTV to get their sports package likely will pay $49 for high-speed Internet access from their cable company, but would only have to pay $39 for that service if they bought it along with cable service.[46]

The price for the bundle, or for add-ons to the bundle, sometimes will vary by customer class, with discounts offered only to new customers or only to customers that the provider is trying to "win back" from another provider or only to some other targeted group of customers. For example, the ILECs are making aggressive efforts to woo back customers with extra incentives, including Visa gift cards and special discounts or credits available only to returning customers.[47] Similarly, cable companies are making aggressive efforts to win back subscription video customers from satellite video providers and high-speed Internet access customers from ILECs. As discussed below in the section on Bundling and Competition, sometimes these efforts can lead to claims that the incumbent is engaging in a price war or even predatory or other anticompetitive pricing behavior.

Table 1. A Sample of Bundled Service Offerings of Major Providers

Package	Components	Price	Availability
AT & T DSL Service with One Rate USA	Unlimited local, local toll, and long distance phone service, and DSL	$89.90 to $94.90*	NJ, NY, MD, MA, VA
Bell South Ultimate Answers	Unlimited local, local toll, and long distance phone service, DSL, and 500 Cingular wireless minutes and 5,000 minutes nights and weekends	$124.98 in GA (prices vary in other states)	Nine states in BellSouth territory
Comcast Corp.'s bundle	High-speed Internet, cable TV	$15 discount on Internet :$42.95 down from $57.95**	35 states plus DC
The Cox Value Bundle	Cable TV, unlimited local and long distance phone service with feature package and high-speed Internet	$120.89	New England
MCI's NeighborhoodHiSpeed	limited local, local toll, and long distance phone service, and DSL	$84.99 to $109.99	29 states plus DC
Owest's Simply Phone Service	Unlimited local and domestic long distance and various premium calling features	$49.99 [add unlimited local wireless for $49.99 or add DSL for $29.99]	Most areas of CO, ID, IA, MN, NE, ND, NM, OR, SD, UT, WA, and WY
SBC Total Connections	Unlimited local, local toll, and long distance phone service, DSL, and 300 anytime wireless minutes [Cingular Wireless] and 5,000 minutes nights and weekends	$90 to $95	AK, CA, CT, IL, KS, MI, MO, NV, OH, OK, TX, WI
Sprint Complete Sense Unlimited with PCS	Unlimited local, local toll, and long distance phone service and unlimited wireless	$179.99 to $189.99	36 states plus DC
Verizon Freedom All	Unlimited local, local toll, and long distance phone service, DSL, and 400 anytime wireless minutes, unlimited nights and weekends and 1,000 mobile-to-mobile minutes	$114.89 to $124.89	MA, NJ, NY, PA, VA
Vonage Premium Unlimited Plan	Unlimited local, local toll, and long distance phone service	34.99***	50 states

Source: Josh Long, "Unwrapping the Bundle: Telcos Tout Retention Factor, But Packages Reduce Profit Margins," [http://www.xchangemag.com/articles/3clcoverstoryl.htm l], viewed on 1/6/2004.

* Excludes $20 discount first three months on DSL; ** excludes the charge for cable television service available at various levels depending on the specific service chosen; *** excludes charges for a telephone line and DSL service or for cable modem service that are needed in order to use Vonage's service.

Bundling, while an effective strategy for reducing churn among high-end customers, also will result in previously full-price customers switching to discount plans, according to Jeffrey Halpern of Sanford C. Bernstein and Co.[48] According to a research report prepared by Roger Sachs, of Cathay Financial, sales of bundled packages have had a mixed impact on the balance sheet: "While churn rates have been reduced, [Bell] profit margins are falling under pressure. In an effort to reduce local churn, SBC has aggressively provided high-speed data and long-distance service at the expense of profitability.[49]

Bundling appears to be primarily a strategy for deterring chum among high-usage customers, at the expense of profit margins. One possible consequence of bundling is that providers will feel the need to buttress their overall profit margins by raising the rates for their stand-alone services, which are the services most frequently purchased by low-usage customers. These customers tend to be less price sensitive than larger users and thus tend to be loyal to their traditional providers. Similarly, low-income and elderly customers are less likely than more affluent and younger customers to seek high-speed Internet access and wireless services, and thus more likely to purchase stand-alone telephone services rather than bundled service offerings.

Public Policy Issues Created by Bundling

Bundling potentially creates several public policy issues for Congress, including the on-going viability of the current funding mechanism for the federal Universal Service Fund, proper treatment of taxes that are assessed on only a subset of services included in a bundled offering, and maintaining competitive markets.

Bundling and the Federal Universal Service Fund

Bundled offerings typically include some combination of interstate telecommunications services, international telecommunications services, intrastate telecommunications services, and non-telecommunications services (information services, such as Internet access, video services, and even customer premises equipment) for a single price. The federal Universal Service Fund - the federal subsidy program that assures affordable telephone rates for high-cost (rural) and low-income telephone customers as well as for schools, libraries, and rural health facilities - is supported by an assessment on interstate and international telecommunications revenues only.[50] But it is difficult to identify the portion of revenues generated by a bundled service offering attributable to the interstate and international telecommunications portion of that bundle. Providers must assign a portion of the bundled price to interstate and international telecommunications services and the fund administrators must be able to audit the attribution to protect against companies gaming the system by understating the interstate and international telecommunications portion. There often is no way, however, to unambiguously assign a portion of the revenues to interstate and international telecommunications, and thus there is uncertainty for both providers and administrators.

This is not a trivial problem. With more than 40% of residential customers now purchasing bundled services (and many business customers obtaining complex bundles of services or bandwidth),[51] it is no longer a simple task to identify interstate and international telecommunications revenues. The federal Universal Service assessment on interstate and international revenues for the first quarter of 2004 is 8.7% [52] Providers usually choose to recover this cost directly from their customers, who would prefer to avoid the assessment. Providers therefore have the incentive to offer their bundled service offerings in a fashion that allows them to minimize the portion of the bundled price attributable to interstate and international telecommunications. Reporting and auditing the interstate and international telecommunications portion of provider revenues is a difficult task.

Table 2. Universal Service Assessment Base: Total Interstate and International End-User Revenues Less Certain Exempt Revenues (in billions)

Year	Total Interstate and International End-User Revenues Less Certain Exempt Revenues
2002	$76.285
2001	$78.461
2000	$78.977

Source: 2002: Federal-State Joint Board Monitoring Report, released December 2003, table 1.9, Preliminary 2002 data, at p. 1-32; 2001: Federal-State Joint Board Monitoring Report, released December 2003, table 1.4, at p. 1-17; 2000: Federal-State Joint Board Monitoring Report, released October 2002, Table 1.4, at p. 1-16

At the same time, as shown in table 2, the Universal Service assessment base - total interstate and international telecommunications end-user revenues less certain exempt international revenues - has been declining as e-mail and instant messaging increasingly substitute for long distance calling and as long distance rates continue to fall.[53] Although data are not yet available for 2003, it is likely that the assessment base continued to decline in 2003 and continues to decline today as far more customers that are high users of interstate and international service have shifted to bundled service offerings with unlimited usage or high usage levels at flat rates that have continued to fall.

As interstate and international telecommunications revenues have begun to fall, and as bundling makes it increasingly difficult to identify and assess those dwindling revenues, many observers are concerned that interstate and international telecommunications revenues no longer provide a sufficient - and sustainable - universal service funding assessment base, as required by the Communications Act.[54] The FCC first issued a Notice of Proposed Rulemaking to address this concern in May 2001,[55] and has subsequently issued additional notices and orders, but to date has taken action only on a few narrow issues. In one action, as mobile wireless telephone service, which typically is offered for a flat rate, has come to be used increasingly for long distance calls, the Commission has increased the "safe harbor" portion of revenues that mobile wireless carriers can attribute to interstate and international calls from 15% to 28.5% [56] The FCC has been partially constrained in its ability to address the issues relating to bundling and the sufficiency of the Universal Service funding mechanism by the language in the Act and by the Fifth Circuit decision [57]. Some parties have claimed that alternative funding mechanisms would not meet the statutory requirements as interpreted by the court, and that Congressional action would be needed to implement these options. [58]

Three options have been proposed to address the issues of bundling and the sufficiency of the Universal Service funding mechanism: expanding the assessment base to include intrastate as well as interstate and international telecommunications services, replacing the current mechanism with a capacity-based assessment on all interstate connections to the public network, and replacing the current mechanism with an assessment on all telephone numbers.

The option to expand the assessment base to include intrastate as well as interstate and international telecommunications services would significantly expand the assessment base, [59] but would only address the reporting and auditing problems created by those bundled service offerings that consist entirely of telecommunications services. It would not address how to attribute the revenues from the interstate, international, and intrastate telecommunications portion of a bundle sold at a flat rate that also includes information services, such as high-speed Internet access, video services, or equipment. Many business customers purchase a fixed amount of bandwidth that they use to provide a wide variety of services - voice, data, Internet access, video conferencing, etc. Currently, many bundled residential offerings that include both telephone service and other services have separate prices for the non-telephone services. But providers will have the incentive to reduce the portion of the bundle subject to the Universal Service assessment on telecommunications services. One way to do this is to simultaneously lower the rate for the telecommunications portion of the bundle and raise the rate for the non-telecommunications portion, but tie the former to the latter so that customers who do not purchase the non-telecommunications services cannot take advantage of the lower telecom rate. Another way to do this is simply to set a single rate for the bundled telecommunications and non-telecommunications services, with the provider determining the portion to be attributed to telecommunications services and thus subject to assessment. Given the Fifth Circuit decision, a statutory change would be required to allow the FCC to impose a Universal Service assessment on intrastate as well as interstate and international telecommunications revenues.

The option to implement a capacity-based assessment on all interstate connections to the public network would avoid the reporting and auditing problem that currently exists for bundled service offerings because it would be based solely on the capacity of each end-user customer connection to the public network. The FCC, with guidance from Congress, could set assessment rates by weighing various public policy considerations to determine, for example, whether, in order to foster broadband deployment, the assessment on high-speed connections (at least for residential customers) should not be set higher than that on standard voice connections, or whether, for equity reasons, there should be a lower assessment for voice grade connections than for high-speed connections. Also, since all end users ultimately must connect to some network to communicate - no matter what technology they use - the assessment base will be sustained over time. A capacity-based assessment on all connections would be simple to implement and administer for residential customers, but far more difficult for business customers, who use many different connection configurations. Some parties have argued that a connection-based approach would require a statutory change because some interstate carriers do not offer connections and thus such a charge would not meet the statutory requirement that all interstate carriers contribute to the fund on an equitable and nondiscriminatory basis.

The option to implement an assessment on all telephone numbers also would avoid the reporting and auditing problem that currently exists for bundled service offerings because the assessment would be based solely on the number of telephone numbers provided to customers. With each telephone number given the same weight, this approach would treat more intensive and less intensive users of the public network exactly the same. But there are so many telephone numbers that the assessment per telephone number is likely to be relatively small. It is possible that a massive move to Internet protocol technology could result in many parties using "addresses" other than the traditional telephone number, but presumably in that case there would be a way to assess the new address; some sort of address will always be needed in order to direct communications from a sender to a receiver. Since some interstate telecommunications carriers do not provide telephone numbers, some parties have argued that a statutory change would be required to implement a telephone number-based universal service assessment mechanism, unless a hybrid assessment mechanism were created that assessed on the basis of revenues those providers of interstate services that do not use numbers. Such a hybrid solution might not eliminate the need for a statutory change, however, if the interstate services provided by those carriers are bundled in a fashion that makes it difficult to identify unambiguously the interstate and international revenues generated.

Currently, no bills have been introduced in Congress that directly addresses the issue of the federal Universal Service assessment base. Both Section 3 of S. 1380, the Rural Universal Service Equity Act of 2003, and Section 4 of H.R. 1582, the Universal Service Fairness Act of 2003, would require the Comptroller General to submit a report to Congress on "the need to reform the high-cost support mechanism for rural, insular, and high cost areas," including a discussion of whether "amendments to section 254 of the Communications Act of 1934 (47 U.S.C. 254) are necessary to preserve and advance universal service." Section 4 of S. 150, the "Internet Tax Non-discrimination Act of 2003, states that "Nothing in the Internet Tax Freedom Act shall prevent the imposition or collection of any fees or charges used to preserve and advance Federal universal service or similar State programs authorized by section 254 of the Communications Act of 1934." Bundling and Taxes.

In addition to the federal Universal Service Fund, there are a number of taxes that are assessed on one or more, but not all, of the services included in various bundled service offerings. This creates the same assessment and auditing problem for these taxes as exists for the federal Universal Service Fund.

In particular, many state and local jurisdictions assess taxes on telephone and/or video services. How should those taxes be assessed on bundled services offered at a flat-rate that include telephone service and high-speed Internet access service or cable service and high-speed Internet access service? How can providers identify and report, and state and local tax collectors audit, the taxable portion of such bundles? The Internet Tax Freedom Act moratorium on taxing Internet access (P.L.107-75) has expired. S. 150, S. 52, and H.R. 49 would make the moratorium permanent; H.R. 1481 would extend the moratorium until 2008. Since it is likely that Internet access services increasingly will be bundled with other services that are subject to local or state taxes, if the Internet tax moratorium is extended the challenge of appropriately assessing and auditing these taxes will grow.

BUNDLING AND COMPETITION

Some observers have been concerned that bundled service offerings could have anticompetitive consequences if they foster industry consolidation or if a provider has market power for one of the services in its bundled offering and can use that offering to tie that service to a competitive service in a fashion that reduces competition for the competitive service.

Although it is too early to determine which providers ultimately will benefit most by the trend toward bundled service offerings, the early market results suggest that the ILECs have been more successful at capturing long distance customers than the long distance companies have been at capturing local customers Despite the fact that the long distance carriers had been able to offer bundles of local and long distance services for months or even years before some of the Regional Bell Operating Companies ("RBOCs") received FCC permission to offer long distance service within their services areas, the RBOCs have captured four long distance customers for every local customer captured by the long distance carriers, as shown in table 3.

According to a J.D. Power and Associates consumer survey, 40% of the respondents stated they would most likely choose their local telephone company to provide bundled services, 21% would most likely choose their long distance company, and 16% their cable company[60]. According to company officials, Verizon now has signed up more than 50% of its local residential customers in some states for long distance service; by contrast, AT & T has signed up at most 15% of its customers to local as well as long distance service.[61] According to Kate Griffin, a senior analyst with the Yankee Group, "The local providers have an edge. The local relationship is worth more. Customers are more likely to look to the local provider for that bundled offering.[62]

Table 3. "Non-traditional" Customers Captured by Local and Long Distance Carriers

Traditional Local Carrier	Long Distance Customers Captured	Traditional Long Distance Carrier	Local Customers Captured
Verizon	15.9 million	AT and T	3.5 million
SBC	11 .5 million	MCI	3.5 million
BellSouth	3.4 million	Sprint	0.2 million*

Source: Griff Witte, "An Evolutionary Edge: Local Phone Firms Pass Long-Distance Companies," Washington Post, December 3, 2003, at p. El. All data provided by the companies. Numbers for traditionally local companies include a limited number of business customers. *Sprint also has 5.3 million local customers in the various territories where it also is the incumbent local exchange carrier.

It may be too soon to conclude how this competition will play out, however. The RBOCs' success may be explained in part by the fact that for more than 20 years residential customers have been choosing among competitive long distance carriers and thus they are not reluctant to switch to their ILEC when that ILEC begins to offer long distance service. On the other hand, residential customers for the first time can choose their local provider and perhaps many simply are not yet ready to change their behavior.

As explained earlier, bundles appeal primarily to heavy telecommunications users. Heavy long distance users already are used to choosing among - and changing - carriers. Heavy local users, however, are just becoming accustomed to choice in local service. Also, local service provides the "lifeline" to the outside and thus customers may tend to be more cautious about leaving their traditional local provider. This pattern may change as customers become used to having local as well as long distance options. Moreover, as local and long distance telephone services become elements of larger bundled offerings that include wireless, video, high-speed Internet access, and other services, the competitive options available to consumers increasingly will come from non-wireline providers.

If there still are impediments to the competitive provision of one of the services included in a bundled offering, however, bundling could allow those providers that are not constrained by those impediments to extend their market advantage beyond the market for that particular service into the markets for the other services included in the bundled offering. According to a study performed by a coalition of small CLECs,[63] the RBOCs have a 61% share of the customers currently using bundled service offerings that include both local and long distance service, and 80% of the remaining customers for bundles that include local and long distance service are served by the unbundled network element known as the platform ("UNE-P") leased from the RBOCs.[64] Under current conditions, if CLECs were to be denied access to UNE-P to offer local residential service, many would not be able to compete with the RBOCs for the provision of bundled local-long distance service. Given the popularity of such bundled services, this likely would allow the RBOCs to extend their advantage in the local market (by dint of their historical position as the monopoly providers with ubiquitous local networks) into the long distance market. Any harm to consumers from lack of competitive choice might be ameliorated, however, by the extent to which other providers could enter to offer bundled local and long distance service. For example, cable companies or other potential competitors could use voice over Internet protocol (VoIP) to offer competitive bundled local-long distance service . Such service may require a customer to have high-speed Internet access, but the high-usage customers most attracted to bundled services often are the consumers most likely to have high-speed Internet access.

A number of CLECs and CLEC customers have brought antitrust suits against RBOCs, alleging that the RBOCs violated the antitrust laws by not making their unbundled network elements available in a timely and viable fashion.[65] The U.S. Supreme Court, however, ruled on January 13, 2004, in Verizon Communications Inc. v. Law Offices of Curtis V. Trinko, LLP,[66] that failure to meet the unbundling requirements in the 1996 Telecommunications Act, which were intended to foster competition by aiding competitors, does not meet the pre-existing antitrust standards, which relate only to acts that would lessen competition, and thus does not represent an antitrust violation. To the extent that access to UNEs are needed for the competitive provision of bundled service offerings, then, oversight can only be performed by federal and state regulatory agencies, not by the antitrust authorities.

Bundling also could affect the competitive environment if it provides a vehicle for a provider with some market power in the market for one of the bundled service elements to price in a fashion that undermines competition in the market for other services in the bundle. Dominant firms typically are constrained in their ability to practice predatory pricing or other potentially anticompetitive types of strategic pricing against new entrants because it is very difficult to introduce selective price cuts for those customers they want to keep away from competitors without giving the same price discount to a large portion of their customer base. This could so dilute revenues and profits in the short term that such losses could not be recouped in the long term even if competitive entry were retarded or entirely eliminated. To the extent a dominant provider attempting to fend off competitive entry could limit the price cuts to those customers most likely to shift providers and to a limited period of time, the potential for anticompetitive predatory or strategic pricing increases. Bundling might be a tool that could facilitate this.

For example, observers have speculated that cable companies, which are the largest providers of high-speed Internet service, might be able to impede ILEC entry into that market by selectively reducing prices for their cable modem services when ILEC entry is imminent . Such concern was kindled recently when Comcast made a targeted e-mail offer to certain customers in California, Maryland, and Illinois, for cable modem service at $19.95 per month for a year.[67] After the promotional period, the price goes back to $42.95. Comcast executive vice president for marketing Dave Watson stated that "This highly targeted e-mail offer is a test campaign aimed directly at DSL customers. It is similar to other win-back-type programs we've conducted in the past. This particular campaign is a limited offer and we anticipate it to be a one-time event as other offers of shorter duration such as 6 months have proven successful." But if Comcast (or any other provider of high-speed Internet access service or any other service that is part of a bundled service offering) has the ability to selectively restrict price cuts to those customers most likely to shift to a competitor and to the time period when a competitor is entering the market, some observers say, then there is at least some potential for that dominant firm to strategically restrict competition in the market even if it is not practicing predatory pricing. The lower prices benefit the selected customers in the short run but can be harmful to the public if they successfully forestall competitive entry. Consumers would then be denied the competition-driven benefits of lower prices and more innovation in the long run.

Another potential competitive consequence of the trend toward bundled service offerings is the incentive created for firms to consolidate in order to more efficiently provide broad bundled offerings or to deny competitors access to independent providers of services needed to offer a complete bundle of services. As explained earlier, consumer preference for larger bundled offerings tends to favor large companies able to offer all or most of the services in the bundle on their own, without reliance on independent entities. But no provider today has the capability of providing all these services. This has fostered marketing agreements and other relationships, which could be a precursor for more formal ownership consolidation. Such consolidation potentially reduces administrative and coordination expenses. But such consolidation also potentially locks up suppliers.

For example, consider the strategy of bundling wireline and wireless service discussed in the section on Wireline, Wireless, and Video Bundling Strategies. The three largest ILECs, Verizon, SBC, and BellSouth, each have large equity interests in wireless carriers, and each have bundled offerings that combine wireless and wireline service . Other wireline carriers face strong market pressure to offer bundles of wireline and wireless services as well. AT & T has a pilot program to offer bundled service in conjunction with AT & T Wireless, which had been part of AT & T but now is independent. AT & T Wireless, however, has announced that it is accepting Cingular's bid to acquire AT & T Wireless. For Cingular, and its parents, SBC and BellSouth, acquisition of AT & T Wireless would provide economies of scale and needed spectrum in large markets such as New York where Cingular has limited spectrum. In addition, such a purchase would take away from AT & T an independent source of wireless service and perhaps make it more difficult for AT & T to offer a bundle that includes wireless service.

Conclusion

The bundling of residential telephone, Internet, and video services has been warmly welcomed by consumers. It allows providers to reduce costly customer churn and exploit marketing efficiencies that they have passed through to consumers by lowering rates. But bundling represents a strategic response to the convergence of previously distinct markets and that convergence is creating the need to review current telecommunications law and rules. Leaders in both the House and the Senate Commerce Committees have indicated that review and reform of the 1996 Telecommunications Act will be on the agenda in the 109 Congress. Major issues that are likely to be addressed include creation of a sufficient and sustainable funding mechanism for the federal Universal Service Fund as interstate telecommunications revenues continue to decline (and become increasingly difficult to identify as bundling proliferates), the proper regulatory treatment of services that are provided by different underlying technologies but compete with one another, and the best regulatory framework for fostering innovation and investment while safeguarding consumers and competition. All three of these issues are likely to be affected by the deployment of Voice over Internet Protocol (VoIP) technology, which already has begun to occur.

References

[1] ILECs are the carriers that were the monopoly providers of retail local telephone service before the 1996 Telecommunications Act opened up local markets to competition.

[2] CLECs are the companies - including the traditional long distance carriers - that began providing local telephone service after the 1996 Act removed statutory prohibitions on competitive provision of local service.

[3] See "Stevens Foresees Telecom Act II in 2005," *Communications Daily*, January 27, 2004, at pp. 1-2. Barton Seeking to Lead House Commerce Committee," Communications Daily, February 5, 2004, at pp. 1-2; and "Analysts Tell House to Expect Dynamic VOID Growth Soon," *Communications Daily*, February 5, 2004, at pp. 5-6.

[4] See Shawn Young, "Phone-Service Bundles Could Backfire as Customers Switch," *Wall Street Journal*, November 7, 2003, at p. B-1.

[5] J.D. Power and Associates 2003 Residential Long Distance Service Study press release, "Customer Satisfaction Increases as Stiff Rate Competition and Bundling Cause Steep Drop in Long Distance Spending," July 1, 2003, at p. 1, posted on [http://www. jdpower.com /news/releases/index .asp].

[6] Griff Witte, "An Evolutionary Edge: Local Phone Firms Pass Long-Distance Companies," Washington Post, December 3, 2003, at p. El, citing Yankee Group senior analyst Kate Griffin.

[7] Ibid at p. E1.

[8] According to the J.D. Power and Associates 2003 Residential Local Telephone Customer Satisfaction Study press release, "Household Switching of Local Service Carriers Increases as New Players Enter the Local Telephone Service Market," July 15, 2003, at p. 1, posted on [http://www.jdpower.com/news/releases/index.asp], "The number of households reporting they have switched local telephone service carriers in the last year has increased more than 60 percent in 2003, rising to 10 percent from 7 percent in 2002." Similarly, according to the J.D. Power and Associates 2003 Residential Cable/Satellite TV Customer Satisfaction Study press release, "Average Monthly Spending for Satellite Service Drops Below Cable Service for the First Time as Cable Market Share Continues to Decline," August 19, 2003, at p. 1, posted on [http://www.jdpower.com/news/releases/index.asp], "Currently, 60 percent of households surveyed subscribe to cable service, down from 68 percent five years ago, while satellite subscriptions have increased from 7 percent of households in 1998 to 17 percent in 2003."

[9] See footnote 12 below.

[10] Glenn Bischoff, "The Principal of Portability," *Wireless Review*, November 1, 2003, at p. 11.

[11] J.D. Power and Associates 2003 U.S. Wireless Regional CSI Study press release, "Customer Loyalty Becoming a More Critical Issue in the Wireless Industry as Phone Number Portability is Poised to Become a Reality in November," September 30, 2003, at p. 1, posted on [http://www.jdpower.com/news/releases/index.asp].

[12] See, for example, Glenn Bischoff, "The Principal of Portability," Wireless Review, November 1, 2003, at p. 11, Tim McElligott, "Churn plus portability equals Y2K-03," *Wireless Review*, September 1, 2003, at p. 35, and Aude Lagorce, "The Battle Over Cell Phone Business Accounts," Forbes.com, http://www.forbes.com/2003/11/13 /ex-al-1113phones.html,viewed on January 6, 2004. The J.D. Power and Associates 2003 U.S. Wireless Regional CSI Study press release, "Customer Loyalty Becoming a More Critical Issue in the Wireless Industry as Phone Number Portability Is Poised to Become a Reality in November," September 30, 2003, at p. 1, posted on http://www.jdpower.com/news/releases/index[.asp], states that 26% of the subscribers in its survey stated they had switched wireless carriers at least once in the past 12 months. Since some of those subscribers may have switched carriers more than once, this suggests a churn rate in excess of 26% .

[13] According to Richard Wolniewicz, "Building a better business one customer at a time," http://telephonyonline.com/ar/telecom [building_better business/], November 12, 2003, (viewed 1/12/04), "Customer churn is one of the most pressing issues the telecommunications industry faces and it affects all types of carriers from cable operators to mobile service providers. According to a study by Bain and Co., companies can boost revenues by as much as 85% if they can retain only 5% more of their best customers."

[14] High churn-related costs may have been responsible for the ineffectiveness of one of the provisions in the 1996 Telecommunications Act intended to foster competitive entry. Section 252(d)(3) requires the ILECs to make their retail services available to new entrants at wholesale rates determined "on the basis of retail rates charged to subscribers for the telecommunications service requested, excluding the portion thereof

attributable to any marketing, billing, collection, and other costs that will be avoided by the local exchange carrier." In implementing this requirement, the FCC adopted a rule instructing states to set wholesale rates by using a methodology that subtracted from retail rates the ILECs' embedded retail-related costs, which were in the range of 15%-20% of retail rates .This is sometimes called the wholesale "discount" off retail rates . With competitive entry, and the resultant customer churn, however, marketing and sales costs for both entrants and incumbents have risen significantly, and far exceed the 15%-20% discount off retail rates . New entrants therefore have not found it viable to enter the market by reselling retail ILEC services at the discounted wholesale rates . The ILECs, on the other hand, have argued successfully in court that the "discount" off the retail price should be reduced, not increased, because some retail costs are fixed, will not decline in proportion to the number of customers lost to the resellers, and therefore "will not be avoided." (Iowa Utilities II, 8th Federal Circuit Court of Appeals, 219 F3.d at 754). The 8th Circuit has remanded the FCC's wholesale pricing rule back to the Commission, which has opened up a proceeding to address that and other cost rules.

[15] Among the many firms that collect data and/or perform chum analysis for providers are Convergys' Knowledge Management Services, Yankee Group, In-Stat/MDR, Zelos Group, iGillottResearch Inc ., Gartner Group, Solomon Wolff Associates, Athene Software, Convergence Consulting Group, Compete Inc ., and Dietrich Lockhard Group.

[16] See, for example, Keith Damsell, "Telecom bundling seen luring customers ; Grouping services together for lower price builds loyalty, trims 'churn,' study says," The Globe and Mail, 29 September 2003, at p. B8, citing Convergence Consulting Group Ltd. study, The Battle for the North American Couch Potato, and referring to Cox Communications' extremely low churn rate with the "triple play" of digital television, high-speed Internet access, and local telephone service.

[17] See Griff Witte, "An Evolutionary Edge: Local Phone Firms Pass Long-Distance Companies," *Washington Post*, December 3, 2003, at p . El.

[18] Vince Vittore and Glenn Bischoff, "Bundling Strategy Provides Soft Landing," *Telephony*, October 7, 2003, at pp . 6-9 .

[19] For example, Covad Communications, which offers high-speed Internet access service using DSL technology, increasingly markets its services through the bundled service offerings of CLECs. See, for example, "AT & T, Covad Extend Consumer Bundle in Midwest States," at [http://www.phoneplusmag.com/hotnews/3chll123256.htm], viewed 1 on 1/7/2004. Covad's marketing partnerships with many CLECs also have been motivated in part by the FCC's Triennial Review Order, 47 CFR Part 51, "Review of the Section 251 Unbundling Obligations of Incumbent Local Exchange Carriers; Implementation of the Local Competition Provisions of the Telecommunications Act of 1996; Deployment of Wireline Services Offering Advanced Telecommunications Capability; Final Rule and Proposed Rule," Federal Register, Vol.68, No. 169, September 2, 2003, at pp. 52276ff. That Order phases out the "line sharing" requirement, under which companies such as Covad could lease from the ILEC, at cost-based rates, the "data" portion of the local loop to a customer's premises to offer that customer high-speed Internet access service while that customer continued to receive voice service from the ILEC. At the same time, the Order imposed a "line splitting" requirement, under which an ILEC must make the local loop available in such a fashion

that a CLEC, such as Covad, that specializes in offering high-speed Internet access (data) service, and another CLEC, such as AT and T or MCI, that specializes in offering voice services, could jointly use the loop to provide the customer both voice and Internet access services. Under line splitting, one of the CLECs leases the entire loop from the ILEC at cost-based rates. The two CLECs then work out between themselves the charges for use of the two (voice and data) portions. The ILEC is merely obligated to make it physically possible for the two CLECs to split the loop, for example, by allowing the CLECs to collocate their equipment with one another within the ILEC's central office.

[20] See, for example, Shawn Young, "Phone-Service Bundles Could Backfire as Customers Switch," Wall StreetJournal, November 7, 2003, at p. Bi, quoting Wayne Huyard, president of mass markets for MCI: "Churn has increased. We are entering an era of commoditization for local and long distance."

[21] UNE-platform consists of the combination of the local loop from the customer premise to the ILEC's central office and the switch port at the central office.

[22] The 1996 Telecommunications Act attempted to foster competitive provision of local telephone service by requiring the ILECs to make available to new entrants those elements of the ILEC networks to which the new entrants needed access in order not to be "impaired" in their ability to offer local service. This requirement that the ILECs unbundlethe elements of their networks and make them available to CLECs should not be confused with the current strategy of many providers to bundleretail services into offerings intended to reduce customer churn.

[23] See "Measuring RBOC Dominance of Bundled-Services: The Progress of Competition Under the New Social Contract," an undated report of the PACE Coalition (a coalition of small CLECs) prepared in late 2003, at p. 3, posted on [www.pacecoalition.com], viewed on 2/10/2004. The FCC's semi-annual report, Local Telephone Competition : Status as of June 30, 2003, which was released in December 2003 and is available at [http://www.fcc.gov/wcb/stats], presents data collected from both ILECs and CLECs for the same period of time. Although it is not possible to directly compare the PACE and FCC data, they appear to describe a consistent scenario. According to the CLEC-provided data presented in Table 3 of the FCC report, in June 2003 58.5% of all CLEC end-user switched access lines were provided using UNEs, 18.2% were provided by reselling ILEC retail services, and 23.3% were CLEC-owned (i.e., self-provisioned). According to the CLEC-provided data presented in Table 2 of the report, 62% of CLEC end-user switched access lines served residential and small business customers and 38% served large business customers. According to the ILEC-provided data presented in Table 4 of the report, 13,026,000 of the 17,231,000 (75.6%) end-user switched access lines that ILECs have provided to CLECs as UNEs were provided as part of UNE-P (as UNEs with switching). Virtually none of the access lines self-provisioned by CLECs serve residential or small business customers (the exception being the unusual case of a residential or small business customer being located on the same site as a large business customer) and virtually none of the CLECs' large business customers are served by resold ILEC retail service (ILEC retail service would rarely meet the needs of a large business customer). Also, those CLEC large business customers not served by CLEC-provisioned loops are far more likely than residential and small business customers to have been served by unbundled loops rather than UNE-P, since it is these large business

customers that CLECs can serve most efficiently with their own switching. The FCC data suggest that approximately 14.7% of CLEC switched access lines used UNEs and served large business customers [38% minus 23.3%]. 44.2% of CLEC switched lines were served by UNE-P [(58.5%)(75.6%)] and 14.3% by UNE-loop [58.5 minus 44.2]. It is likely that virtually all of the 14.3% of CLEC UNE-loop switched access lines served large business customers, since it is far more efficient to serve these customers than residential and small business customers with UNE-loop. Thus, only about 0.4% of CLEC switched access lines used UNE-platform and served large business customers [14.7% minus 14.3%]. This suggests that 43.8% of CLEC switched access lines served residential and small business customers using UNE-P, which would be 70% of the 62% of CLEC switched access lines serving residential and small business customers. While this number may be slightly overstated because it implicitly assumes no residential and small business customers are served by UNE-loops, it is not out of line with the 80% PACE findings. The FCC data cover all CLEC lines; the PACE data cover only those lines sold as part of a bundled local-long distance service offering. While virtually all CLECs that use UNEs are offering such bundles, some of the cable companies, which do not use UNEs, offer local telephone service but not bundled with long distance service. Thus one would expect the figure constructed from the various FCC tables (70%) to be lower than the PACE figure (80%).

[24] ILECs are not required to offer voice mail because it is an information service, not a telecommunications service, and thus not subject to the 1996 Act's unbundling requirements. Similarly, privacy management services are provided through the ILECs' Advanced Intelligent Networks (AIN), which are proprietary and which the ILECs are not required to make available to CLECs.

[25] To offer its own high-speed Internet service to a customer over ILEC lines, a CLEC must collocate its own DSL equipment (DSLAMs) at a particular point in the ILEC's network . It is only viable to deploy such equipment in places where the CLEC can expect to capture enough customers for the DSL service to justify the investment and where there is space at the ILEC location to place that equipment .

[26] *Communications Daily*, January 8, 2004, at p. 6.

[27] In its September 2003 Triennial Review Order, the FCC ended the "line sharing" requirement that ILECs make the data portion of their local loops available to data CLECs at cost-based rates. However, the FCC has grandfathered cost-based prices for the data portion of the customer line for those customers who were served by the high-speed Internet access service provider prior to that Order.

[28] See footnote 19 above.

[29] "SBC Communications Provides Progress Report on Major Growth Strategies, Outlines Broad Service and Cost Initiatives," SBC-Press Room, November 13, 2003, http://www.sbc[.com/gen/press-room?pid=4800 and cdvn=news and newsarticleid =2072 1], viewed on 12/05/2003. This statement also appears at "SBC Provides 2004 Outlook, Updates Major Trends,"[http://convergedigest.com/ Bandwidth/sample category article.asp?ID=9406], viewed on 1/9/2004.

[30] Under the terms of the 1996 Telecommunications Act, the Regional Bell Operating Companies (the old Bell System portions of Verizon, SBC, BellSouth, and Qwest) were required to pass a 14 point checklist demonstrating that their markets have been opened to competitive provision of local telephone service before they were allowed to enter

the long haul long distance market in their respective service areas . They have now been approved to offer long distance service in all states.

[31] Griff Witte, "An Evolutionary Edge : Local Phone Firms Pass Long-Distance Companies," *Washington Post*, December 3, 2003, at p . El .

[32] Mike Farrell, "DBS Pitches: MSOs Swine, Telcos Divine," Multichannel News, November 24, 2003, at p. 1, and "BellSouth announces new options," *Alexandria Daily Town Talk*, December 15, 2003, at p. 8B.

[33] Mike Farrell, "DBS Pitches: MSOs Swine, Telcos Divine," *Multichannel News*, November 24, 2003, at p. 1.

[34] National Cable and Telecommunications Association, http://www.ncta [com/Docs/Page Content.cfm?pagelD=9 3], viewed on 1/12/04.

[35] National Cable and Telecommunications Association, h [ttp://www.ncta.com/Docs/ PageContent.cfm?pagelD=32], viewed on 1/12/04.

[36] See Matt Richtel, "Time Warner Deal Raises Ante in Cable's Bid for Phone Market," New York Times, December 9, 2003.

[37] According to the J.D. Power and Associates 2003 Residential Cable/Satellite TV Customer Satisfaction Study press release, "Average Monthly Spending for Satellite TV Service Drops Below Cable Service for the First Time as Cable Market Share Continues to Decline," August 19, 2003, at p .2, posted on [http://www.jdpower.com /news/releases/index.asp], "One area where cable providers may have an opportunity to stem this migration to satellite is in bundling telephony and Internet access with cable TV service . With growing consumer desire to combine multiple services in a single bill for convenience and simplicity, the study finds that 34 percent of cable subscribers want to combine their cable service with some other telecommunications product or service."

[38] See Matt Stump, "Cable Ops Touting VOD As Anti-Chum Weapon," Broadband Week, March 4, 2003, viewed on [h ttp ://www .broadbandweek.co m /news/020304/020304 cable_cableops.htm] on 1/12/04.

[39] See Shawn Young, "Phone-Service Bundles Could Backfire as Customers Switch," *Wall Street Journal*, November 7, 2003, at p. B1.

[40] Ibid at p. Bl.

[41] "Cingular to Acquire AT and T Wireless, Create Nation's Premier Carrier," February 17, 2004, at [http://www.attwireless.com/ press/releases/200 4 releases/021704.ihtml], viewed on 2/17/04.

[42] Cingular succeeded in an informal auction process created when AT & T Wireless announced in January 2004 that its board of directors had authorized the company to entertain acquisition offers after receiving overtures from nearly half a dozen suitors, including Cingular, Nextel Communications, Vodafone, NIT DoCoMo, and AT & T. See Matt Richtel and Andrew Ross Sorkin, "AT & T Wireless for Sale as a Shakeout Starts," *New York Times*, January 21, 2004, at p. Cl. According to the article, the "move by AT & T Wireless and its potential buyers indicates that one of the nation's most fiercely competitive industries is heading toward a long-awaited consolidation that may be the tip of a multilayered and complex merger process around the world."

[43] Shawn Young, "Phone-Service Bundles Could Backfire as Customers Switch," *Wall StreetJournal*, November 7, 2003, at p. Bl.

[44] Ibid at p. BI. The high chum rates that MCI and other CLECs are experiencing with their bundled services (Adam Quinton, a telecommunications analyst at Merrill Lynch, estimates that turnover in bundled plans offered by rivals to the Bell Operating Companies is as high as 8% a month - or nearly 100% in a year- in some highly competitive areas) also is the result of aggressive ILEC campaigns to "win-back" customers lost to the CLECs and to the "sticker shock" customers experience because advertised rates typically exclude fees and taxes that can add as much as 15% to the customer bill.

[45] Geraldine Fabrikant, "In Fight Between Cable and Satellite, Customers Gain an Edge," New York Times, December 1, 2003, at p. C22.

[46] Ibid at p. C22.

[47] Shawn Young, "Phone-Service Bundles Could Backfire as Customers Switch," *Wall Street Journal*, November 7, 2003, at p . BI .

[48] fbid at p. 131.

[49] See Josh Long, "Unwrapping the Bundle : Telcos Tout Retention Factor, But Packages Reduce Profit Margins," [http://www.xchangemag. com/articles13clcoverstoryl .htm 1], viewed on 1/6/2004.

[50] The Communications Act, as amended, in Section 254(d) requires "Every telecommunications carrier that provides interstate telecommunications services shall contribute, on an equitable and nondiscriminatory basis, to the specific, predictable, and sufficient mechanisms established by the Commission to preserve and advance universal service." 47 U.S.C. § 254(d). In Texas Office of Public Utility Counsel v. FCC, 183 F.3d 393 (5 i Cit 0. 1999), the Fifth Circuit overturned an FCC order assessing intrastate as well as interstate telecommunications to fund the schools and libraries portion of the federal Universal Service Fund, but upheld assessing international telecommunications revenues.

[51] According to the J.D. Power and Associates 2003 Major Provider Business Telecommunications Services Study press release, "More Business Data Customers Willing to Switch Telecommunications Providers in Favor of Cost-Saving Bundles," October 2, 2003, at p. 1, posted on [http://www.jdpower.com/news/releases/index.asp], "The study finds that the intention to switch providers to bundle multiple telecommunications services is up 14 percentage points among broadband business customers to 43 percent overall."

[52] "Proposed First Quarter 2004 Universal Service Contribution Factor," FCC Public Notice, released December 4, 2003, at p . 1.

[53] The universal service assessment base is total interstate and international end-user revenues less three exempt categories : revenues for international-to-international service, international revenues where interstate toll represents less than 8% of the company's combined interstate and international revenues, and interstate and international revenues for 2,570 filers who are de minimis and thus not required to contribute . These three categories of exemptions represent approximately 3% of total interstate and international end-user telecommunications revenues .

[54] Section 254(b)(5) of the Communications Act, as amended, lists as a principle that "There should be specific, predictable and sufficient Federal and State mechanisms to preserve and advance universal service." 47 U.S.C. §.

[55] In the Matter of Federal-State Joint Board on Universal Service, Notice of Proposed Rulemaking, CC Docket No. 96-45, released May 8, 2001.

[56] "In the Matter ofFederal-StateJointBoard on Universal Service, Report and Order and Second Further Notice of Proposed Rulemaking, CC Docket No. 96-45, released December 13, 2002, at p. 14.

[57] See footnote 50 above.

[58] See, for example, the Comments filed on June 25, 2001 by Verizon and by the United States Telephone Association, responding to the FCC's May 8, 2001 Notice of Proposed Rulemaking in ln the Matter of Federal-State Joint Board on Universal Service, CC Docket No. 96-45.

[59] Total End-User Revenues from local service, wireless service, and toll service in 2001, the latest year for which data are available, were $235.5 billion, $155.3 billion of which were intrastate, according to the FCC's Annual Trends in Telephone Service, Table 15.1, "Telecommunications Industry Revenues: 2001," (released inAugust 2003) at p.15-3. This report is available at [www.fcc.gov/wcb/stats].

[60] J.D. Power and Associates 2003 Residential Long Distance Service Study press release, "Customer Satisfaction Increases as Stiff Rate Competition and Bundling Cause Steep Drop in Long Distance Spending," July 1, 2003, at pp. 1-2, posted on [http://www.jdpower.com/news/releases/index.asp].

[61] Griff Witte, "An Evolutionary Edge; Local Phone Firms Pass Long-Distance Companies," *Washington Post*, December 3, 2003, at p. El.

[62] Ibid at p. El.

[63] "Measuring RBOC Dominance of Bundled-Services: The Progress of Competition Under the New Social Contract," an undated report of the PACE Coalition (a coalition of small CLECs) prepared in late 2003, at p. 3, posted on [www.pacecoalition.com], viewed on 2/10/2004.

[64] The PACE report claims that only 1% of these customers are served by wireless providers. This suggests that the PACE Coalition only included those wireless customers who have abandoned wireline service entirely and use wireless service as their exclusive provider of local and long distance service.

[65] For example, Trinko and Cavalier brought cases against Verizon, Covad brought a case against BellSouth, and Metronet brought a case against Qwest.

[66] Verizon Communications Inc., Petitioner v. Law Offices of Curtis V. Trinko, LLP, Supreme Court of the United States, 540 U .S. (2004), January 13, 2004, Slip opinion at 6-7 (2004 WL 51011). For a more complete discussion of the Trinko decision, see CRS Report RS21723, Verizon Communications, Inc. v. Trinko: Telecommunications Consumers Cannot Use Antitrust Laws to Remedy Access Violations of Telecommunications Act.

[67] *Communications Daily*, November 17, 2003, at p. 7.

In: Computer Networks, Policies and Applications
Editors: C. M. Berger and A. D. Streusse
ISBN: 978-1-61209-088-7

Chapter 6

THE POSITION OF DIRECTOR OF NATIONAL INTELLIGENCE: ISSUES FOR CONGRESS*

Alfred Cumming

ABSTRACT

The 9/11 Commission, in its recent report on the attacks of September 11, 2001, criticized the U.S. Intelligence Community's (IC) fragmented management structure and questioned whether the U.S. government, and the IC, in particular, is organized adequately to direct resources and build the intelligence capabilities that the U.S. will need to counter terrorism, and to address the broader range of national security challenges in the decades ahead.

The Commission made a number of recommendations, one of which was to replace the current position of Director of Central Intelligence with a National Intelligence Director (NID) who would oversee national intelligence centers on specific subjects of interest – including a National Counterterrorism Center (NCTC) – across the U.S. government, manage the national intelligence program, oversee the agencies that contribute to it, and have hiring, firing and budgetary authority over the IC's 15 agencies. The Commission recommended that the director, and the NCTC, be located in the Executive Office of the President and that a deputy NID be established to oversee the day-to-day operations of the Central Intelligence Agency (CIA). The Commission's recommendation to strengthen management authority over the IC is the latest contribution to an IC structural reform debate that dates at least to 1955, when arguments for stronger IC authority began to surface. OMB deputy director James Schlesinger in 1971 first broached the DNI concept.

Recently some Members of Congress have introduced intelligence community reform legislation that would establish the position of Director of National Intelligence (DNI), or strengthen DCI authorities.

Reactions to the concept of a DNI have been mixed since its inception. Supporters argue that the DCI cannot manage the IC, the CIA and serve as the President's chief intelligence advisor, and do justice to any of the jobs. Other than the CIA, the DCI also lacks hiring, firing and budget authority. They argue that the absence of strong,

* This is an edited, reformatted and augmented edition of a United States Congressional Research Service publication, Report RL32506, dated August 12, 2004.

centralized leadership has resulted in divided management of intelligence capabilities; lack of common standards and practices across the foreign-domestic intelligence divide; structural barriers that undermine the performance of joint intelligence work; and a weak capacity to set priorities and move resources.

Opponents counter that a DNI would lose day-to-day control over the CIA, a natural power base and, as a result, influence. They also contend that placing the intelligence director in the Executive Office of the President, as the 9/11 Commission has proposed, risks the politicization of intelligence, giving the White House more direct control over covert operations, blurring the line between foreign and domestic operations and possibly shifting too much influence over intelligence to the Department of Defense. With regard to DOD influence, other opponents argue that a national director will shift the balance of control away from DOD, risking intelligence support to the warfighter. The congressional role includes deciding whether to establish the position of the DNI and its authority.

Proposals to reorganize the United States Intelligence Community began to surface almost as soon as the management structure for the Intelligence Community (IC) was statutorily established by the passage of the National Security Act of 1947. Since then, at least 19 commissions, committees and panels, created by either the executive or legislative branches, have made numerous recommendations for structural reorganization. Several of the proposals urged stronger centralized IC authority and, in some cases, the establishment of the position of Director of National Intelligence (DNI).

The so-called Second Hoover Commission, established by law to examine the organization of the executive branch in 1953, became the first independent panel to push for stronger centralized IC authority. [See page 15 for a historical review of efforts to centralize and strengthen IC leadership]. In examining ways to strengthen the IC, the commission recommended that an "executive officer" be named to manage the CIA so that the DCI could focus attention on the IC.

Eighteen years and several commissions, committees, and panels after former President Herbert Hoover made his recommendations, the Office of Management and Budget's (OMB) Deputy Director James R. Schlesinger (later a DCI), after studying the IC's management at the behest of former President Richard Nixon, blamed the absence of strong central IC leadership for "unproductively duplicative" intelligence collection systems, and the failure to coordinate the allocation of resources. Schlesinger considered the establishment of a DNI, but backed away, recommending, instead, "a strong DCI who could bring intelligence to an adequate level of quality and responsiveness."[1]

In the most recent iteration of the DNI debate, the 9/11 Commission[2] on July 22, 2004, recommended that the position of the DCI should be replaced by a National Intelligence Director (NID)[3] who would oversee national intelligence centers on specific subjects of interest across the U.S. government, and would manage the U.S. national intelligence program and oversee the agencies that contribute to it. The Commission recommended the establishment of the position of deputy NID for Foreign Intelligence to direct the CIA's day-to-day operations. The Commission also recommended the creation of a National Counterterrorism Center (NCTC), which would be placed in the Executive Office of the President and be under the control of the NID. The NCTC would be a center for joint operational planning and joint intelligence.[4]

Congressional Proposals Concerning DNI Position

There have been a variety of proposals concerning the DNI position. Following the creation of the intelligence oversight committees in the Senate (1976) and in the House (1977), Congress considered charter legislation that included, among other proposals, one that would have created the position of a DNI to manage the IC. A presidentially selected deputy would have managed the CIA. Confronted by strong opposition to the overall legislation, which also included language governing covert actions, the Committees did not report the respective bills.

In 1992, Senator David Boren and Representative David McCurdy, respective chairmen of the Senate Select Committee on Intelligence (SSCI) and House Permanent Select Committee on Intelligence (HPSCI), introduced legislation creating the DNI position and giving the position the authority to program and reprogram funds. Their legislation also would have created a separate director of CIA. Boren and McCurdy failed to win adoption of their legislation in the face of opposition by the Department of Defense (DOD) and the congressional Armed Services Committees.

In December 2002, the Congressional Joint Inquiry Into Intelligence Community Activities Before and After the Terrorist Attacks of September 11, 2001[5], recommended that a new cabinet level Senate-confirmed DNI position be established, and that a separate director be named to manage the CIA. The Joint Inquiry also recommended that the DNI be granted the full range of management, budgetary and personnel responsibilities needed to make the entire IC operate as a coherent whole.[6]

DCI Budget-Related Responsibilities and Authorities under the National Security Act of 1947 Are Seen by Some as Limited

Proponents of establishing the DNI position contend the current IC management structure characterized by an incoherence they attribute to two flaws. First, because the DCI is dual-hatted, heading both the IC and the CIA, they maintain he is too busy to do either job well. Second, they argue that the DCI's hiring, firing and budget authorities are limited. From their perspective, the result is an IC management structure that lacks direction and focus.

Any discussion of the DNI concept invariably leads to a debate over whether the two jobs should be split, and whether current DCI budget authorities are strong enough to permit effective management of the IC.

With regard to budget authority, the National Security Act of 1947 authorizes the DCI to facilitate the development of an annual intelligence budget [1947 National Security Act, see Sec.103.(c)(1)(A) [50 U.S.C. 403-3(c)(1)(A)]. The act also stipulates that the DCI prepare and approve all budgets for each of the IC agencies comprising the National Foreign Intelligence Program (NFIP)[7] [1947 National Security Act, see Sec.104.[50 U.S.C. 403-4] (b)].

More than 85% of the NFIP resources, however, are estimated to have been appropriated to agencies other than the CIA, and remain beyond the DCI's direct control.[8] The lion's share of that amount is appropriated directly to the Secretary of Defense, who then determines how the Pentagon will fund the NFIP's three largest agencies — the National Security Agency (NSA), the National Reconnaissance Office (NRO) and the National Geospatial-

Intelligence Agency (NGA).[9] Each agency, among other tasks, supports Department of Defense (DoD) combat operations, and each answers to the Secretary of Defense. The DCI is left with exclusive budget authority only over the CIA.[10] Some have asserted that, in part because of the DCI's relatively weak position with respect to the IC, DCIs historically have devoted the bulk of their time to managing the CIA and serving as the President's intelligence advisor, rather than overseeing the IC.[11]

Recent Interest in the Establishment of a DNI

During the last two sessions of Congress — the 107th and 108th — Senators Dianne Feinstein, Tom Daschle and Bob Graham, and Representative Jane Harman have separately introduced legislation that would establish the DNI position. Representative Porter Goss also has introduced related legislation, but rather than establishing the position of DNI, his bill would strengthen DCI management authorities with regard to budget and personnel responsibilities.

Senator Dianne Feinstein

Senator Feinstein first introduced DNI-specific legislation (S. 2645, introduced on June 19, 2002) during the107th Congress. Her bill established the position of DNI and authorized a ten-year term for the position. The bill also established a separate position of the director of the CIA, and prohibited the DCI from simultaneously serving as DNI.

Senator Tom Daschle

In the 108th Congress, Senator Tom Daschle was the first to introduce DNI legislation (S. 6, introduced on January 7, 2003) as part of a broader security reform package. His DNI language is almost identical to that included in the Feinstein legislation. The Daschle bill would create the DNI and authorize a ten-year term. Like the earlier Feinstein legislation, Daschle's bill also would establish a separate position of DCI. The two bills differ in one aspect. In contrast to the Feinstein bill, the Daschle legislation would provide the DNI enhanced budget responsibilities over those enjoyed by the DCI under the current National Security Act of 1947.

Senator Dianne Feinstein, Second Proposal

Following the Daschle's bill introduction, Senator Feinstein introduced a revision of her earlier DNI legislation (S. 190, introduced on January 17, 2004). Her new version would eliminate the earlier 10-year DNI term contained in her original bill, and would enhance the DNI's budget responsibilities as the Daschle bill proposed.

Senator Bob Graham

Later in 2003, Senator Graham of Florida introduced DNI legislation as (S. 1520, introduced on July 31, 2003) part of a broader intelligence reform legislative package.[12] His DNI legislative language is identical that contained in Senator Feinstein's S. 190.

Representative Jane Harman

Representative Jane Harman introduced her version of DNI legislation (H.R. 4104, introduced on April 1, 2004) as part of a broader intelligence reform package. Her bill would provide the DNI limited budget authority enhancements over what the DCI currently enjoys.

Representative Porter Goss

Finally, Representative Porter Goss introduced IC organization reform legislation (H.R. 4584, introduced on June 16, 2004) that would strengthen DCI authorities over the IC, but leave the DCI in charge of both the IC and the CIA. Among other changes, it would repeal the current prohibition pertaining to the DCI exercising internal security functions.

What the Bills Are Designed to Accomplish

Feinstein and Graham Would Establish a DNI

The Feinstein and Graham bills would establish a presidentially-nominated, Senate-confirmed position of DNI, who would serve as the head of the IC's 15 separate intelligence agencies, including the CIA.

Both bills would establish a separate Senate-confirmed DCI, who would manage the CIA. Each bill would also prohibit the DCI from serving simultaneously as DNI, and each contains Sense of Congress language indicating that the DNI should be a member of the president's cabinet.

Both bills would establish the Senate-confirmed position of Deputy DNI.

Neither bill contains the ten-year DNI term of service included in Senator Feinstein's original DNI bill (S. 2645).

Feinstein and Graham Would Enhance DNI Budget-Related Responsibilities[13]

The Feinstein and Graham bills would, to some degree, strengthen the DNI's hand in the area of budget responsibilities by making three modifications to the 1947 National Security Act governing current DCI budget responsibilities.

First, both bills stipulate that the DNI shall "develop" an annual intelligence budget [in Feinstein, see Title 2, Sec. 103(b)(1)(A); in Graham, see Sec.2(a)Sec.103(b)(1)]. Under current statute, the DCI's authority is less definitive; he is authorized to "facilitate the development of an annual budget for intelligence...." [in the 1947 National Security Act, see Sec.103(c)(1)(a)].

Second, each bill identically stipulates that the DNI's development of an annual budget shall include the "review, approval, and modification of the execution of intelligence community budgets, and personnel and resource allocation in furtherance of such annual budget." [in Feinstein, see Title 2, Sec.103(b)(1)(A); in Graham, see Sec.2 (a)Sec.103(b)(1)(A)]. The 1947 National Security Act contains no similar language.

Third, both bills identically stipulate that the DNI's development of an annual budget shall include "managing and overseeing the execution and, if necessary, the modification of the annual budget for the National Foreign Intelligence Program (NFIP), including directing the transfer of funds of personnel between elements of the intelligence community." [in

Feinstein, see Title 2, Sec.103 (b)(1)(C); in Graham, see Sec.2(a)Sec.103(b)(1)(c)]. The National Security Act of 1947 contains no similar language.

Feinstein and Graham Would Strengthen DNI Budget and Personnel-Related Authorities

In the area of budget and personnel-related authorities, both bills would strengthen the DNI's budget and personnel authorities by making three changes in the authorities currently enjoyed by the DCI under the 1947 National Security Act.

First, each bill would accord the DNI the authority to transfer personnel during the year of budget execution[14] without being required to develop with the heads of affected departments and agencies the procedures governing such transfers [in Feinstein, see Title 2, Sec.103A.(d); in Graham, see Sec.2(a)Sec. 103A.(d)]. In contrast, the 1947 National Security Act authorizes the DCI to transfer personnel during the year of budget execution, but only "in accordance with procedures to be developed by the Director and the heads of affected departments and agencies..."[1947 National Security Act, see Sec.104 [50 U.S.C. 403-4] (d)].

Second, both the Feinstein and Graham bills would authorize the DNI to transfer funds and personnel from the Federal Bureau of Investigation (FBI). The 1947 National Security Act prohibits such transfers [1947 National Security Act, see Sec. 104 [50 U.S.C. 403-4](d)(2)(A)(iv)].

Finally, both bills would permit the DNI to unilaterally transfer funds and personnel, even if the heads of the affected agencies and departments objected. Under the 1947 National Security Act, department heads are authorized to prevent DCI-initiated fund and personnel transfers by submitting their objections in writing to the DCI [1947 National Security Act, see Sec.104. [50 U.S.C. 403-4](d)(2)(A)(v) and (B)(iii)].

Both bills would mirror the 1947 National Security Act in authorizing DNI approval of all IC component budgets, and granting the DNI veto authority over any NFIP reprogramming.

Feinstein and Graham Bills Silent on DNI Personnel Policies

The Feinstein and Graham bills would not provide the DNI any enhanced hiring and firing authorities beyond those currently enjoyed by the DCI. Those limited authorities do not require the DCI's concurrence on some IC appointments. Consultation with the DCI is required on other IC appointments [1947 National Security Act, see Sec.106[50U.S.C.403-6](a) and (b)].

Daschle DNI Legislation Tracks Feinstein and Graham

Senator Daschle's DNI legislation is identical to that contained in the Feinstein and Graham bills, with one exception; Daschle would establish a 10-year term for the DNI.

Harman Would Establish a DNI

Like the Feinstein, Graham and Daschle bills, Representative Harman's legislation contains language that would establish a presidentially-nominated, Senate confirmed position of DNI who would serve as the head of the IC's 15 separate intelligence agencies, including the CIA.

Mirroring the Feinstein, Graham and Daschle bills, Harman's legislation would establish a separate Senate-confirmed DCI, who would manage the CIA. Her bill also is identical in that it would prohibit the DCI from serving simultaneously as DNI.

Similarly, the Harman bill would establish the Senate-confirmed position of Deputy DNI. In contrast to her colleagues, her legislation proposes that the Deputy DNI serve concurrently as Undersecretary of Defense for Intelligence. Harman's language differs from the other bills in two other respects. First, her bill does not include sense of the Congress language that the DCI should be a cabinet member; second, it would establish a ten-year term for the DCI.

The Harman bill is identical in providing that the DNI act as the principal advisor to the President for intelligence matters, a role now performed by the DCI.

Harman Enhances DNI Budget-Related Responsibilities

The Harman bill would strengthen the DNI's hand in the area of budget-related responsibilities over those currently enjoyed by the DCI, but to a lesser extent than would the Feinsten/Graham/Daschle bills.

The Harman bill tracks the Feinstein/Graham/Daschle bills by stipulating that the DNI shall "develop" an annual intelligence budget [in Harman, see Title 1, Sec. 103(b)(1)]. Under current statute, the DCI's authority is less definitive in that he is authorized to "facilitate the development of an annual budget for intelligence..." [in the 1947 National Security Act, see Sec.103(c)(1)(a)].

But the Harman legislation does not contain the language included in her colleagues' bills authorizing the DNI to:

> review, approval, and modification of the execution of intelligence community budgets, and personnel and resource allocation in furtherance of such annual budget.

and calling on the DCI to develop an annual intelligence budget by:

> managing and overseeing the execution and, if necessary, the modification of the annual budget for the National Foreign Intelligence Program (NFIP), including directing the transfer of funds of personnel between elements of the intelligence community.

Harman Would Strengthen DNI Budget and Personnel-Related Authorities

The Harman bill is identical to the Feinstein/Graham/Daschle bills in authorizing the DNI to transfer personnel during the year of budget execution without having to develop procedures governing such transfers with the heads of the affected departments and agencies [in Harman, see Title I, Sec. 103A.(d)]. Under current statute, the DCI is authorized to transfer personnel during the year of budget execution, but only "in accordance with procedures to be developed by the Director and the heads of affected departments and agencies..." [1947 National Security Act, see Sec.104 [50 U.S.C. 403-4](d)].

The Harman bill, like the Feinstein/Graham/Daschle bills, but unlike the 1947 National Security Act, would authorize the DNI to transfer funds or personnel from the FBI during the year of budget execution.

In contrast, however, the Harman would withhold from the DNI unilateral authority to transfer funds or personnel during the year of budget execution. Rather, the Harman language would provide that the President would arbitrate any disagreement between the DNI and an agency head over fund or personnel transfers, provided the agency head submitted his objection to the President in writing.

Under current statute, an agency head has the authority to prevent DCI-initiated transfers, provided he informs the DCI of his objection in writing [1947 National Security Act, see Sec.104.[50 U.S.C.403-4](d)(2)(A)(v) and (B)(iii)].

The Harman bill, like the Feinstein/Graham/Daschle bills, mirrors the 1947 National Security Act in authorizing DNI approval of all IC component budgets, and granting the DNI veto authority over any NFIP reprogramming.

Harman Would Strengthen other Personnel Authorities

The Harman bill would not provide the DNI hiring and firing authorities over the agency heads of the IC agencies, but would make three modifications in the 1947 National Security Act to provide the DNI with somewhat strengthened personnel authorities over those currently enjoyed by the DCI.

First, the Harman bill would require that the DNI and Secretary of Defense jointly recommend to the President appointments to the positions of directors of NSA, NRO and NGA, respectively [in Harman, see Title XI, Sec.404(a). In contrast, under the 1947 National Security Act, the Secretary of Defense is required to seek the DCI's concurrence before recommending to the President appointments to these positions. The Secretary may make the recommendation to the President if the DCI does not concur, but notify the President of the DCI's non-concurrence [1947 National Security Act, see Sec106[50 U.S.C. 403-6](a)].

Second, the Harman bill authorizes the DNI to recommend to the President an individual for appointment to the position of DCI [in Harman, see Title IX, Sec.4049(b)]. The 1947 National Security Act contains no similar provision.

Third, the bill would require that the DNI and affected departments must *concur* in recommending to the President individuals to fill the following the positions:

- Director of the Defense Intelligence Agency
- The Assistant Secretary of State for Intelligence and Research
- The Director of the Office of Intelligence of the Department of Energy
- The Director of the Office of Counterintelligence of the Department of Energy
- The Assistant Secretary for Intelligence and Analysis of the Department of the Treasury
- The Under Secretary for Information Analysis and Infrastructure Protection of the Department of Homeland Security.

If the DNI does not concur, the department head must notify the President of the non-concurrence. Under current statute, department heads are required to only *consult* with the DCI regarding appointments to these positions [the 1947 National Security Act, see Sec.106.[50U.S.C.403-6](b)].

Fourth, the Harman bill stipulates that the Director of the Federal Bureau of Investigation *must obtain* the concurrence of the DNI before recommending to the Attorney General individuals to fill the following positions:

- The Executive Assistant Director for Intelligence of the Federal Bureau of Investigation
- The FBI's Executive Assistant Director for Counter-Terrorism and Counter-Intelligence

Under current statute, the FBI director is required to provide the DCI *timely notice* of his recommendation to the Attorney General, so that the DCI may consult with the FBI director before the Attorney General appoints in individual to fill the vacancy.

The Harman bill omits the DCI's current authority to hire and fire CIA employees [in the 1947 National Security Act, see Sec.104[50U.S.C.403-4](h).

Goss Legislation Would Not Establish DNI, but Instead Strengthen DCI

The Goss bill would strengthen the DCI's authority over the IC, in part by establishing eight new associate directors and six assistant directors through which the DCI would manage the IC.

Goss Would Appropriate NFIP Funds Directly to the DCI

Perhaps in its most significant change, the Goss bill would appropriate all NFIP funds directly to the DCI [H.R. 4584, see Title I, Sec.107(a)(50(A). Under the 1947 National Security Act, NFIP funds are appropriated directly to individual IC agencies.

Goss Would Strengthen DCI Budget-Related Responsibilities

Rather than authorize DNI to "develop" an annual intelligence budget, as the Feinstein/Graham/Daschle/Harman bills would, or "facilitate the development of an annual budget"as stipulated by the 1947 National Security Act, the Goss bill would authorize the DCI to "determine the" annual budget [Goss, See Title III, Sec. 301(a)].

Goss Would Enhance DCI Budget-Related Authorities

The Goss legislation would strengthen current DCI budget-related authorities in three ways.

First, with regard to approval of IC budgets, the Goss bill stipulates that the DCI, through the newly created positions of Associate and Assistant Directors of Central Intelligence would "direct, coordinate, and prepare the annual budgets of the elements of the intelligence community within the NFIP, in consultation with the heads of those elements" [H.R. 4584, see Title III, Sec.302(b)].

Second, the Goss bill stipulates that the DCI would provide budget guidance for elements not falling under the NFIP i.e. the Joint Military Intelligence Program (JMIP)[15] and the Tactical Intelligence and Related Activities (TIARA)[16], both directly controlled by the

Secretary of Defense [H.R. 4584, see Title III, Sec.302(b)]. There is no similar language in current statute or the Feinstein/Graham/Daschle/Harman bills.

Third, the Goss bill appears to strengthen the DCI authority over budget approval, stating that IC budgets "*may not* be provided to the President for transmission to Congress unless the Director has approved such budget." [Title III Sec. 302(b) (4)]. In contrast, the Feinstein/Graham/Daschle/Harman bills contain language identical to that in current statute, which stipulates that the DCI shall "approve such budgets before their incorporation in the National Foreign intelligence Program" [1947 National Security Act, Sec.104 [50 U.S.C. 403-4] (b)].

Goss Would Strengthen DCI Reprogramming Authorities

Whereas the Feinstein/Graham/Daschle/Harman bills do not address DNI reprogramming authorities, the Goss bill does, by strengthening the DCI's hand to reprogram funds. The current statute states that no NFIP funds may be reprogrammed without the DCI's prior approval, "except in accordance with procedures issued by the Director."[Sec. 104. [50 U.S.C. 403-4] (c)]. The Goss bill eliminates the procedures requirement, effectively ruling out any exceptions to the DCI's reprogramming authority.

Goss Would Strengthen DCI Budget and Personnel Transfer Authorities

The Goss bill would make three modifications to current statute with regard to budget and personnel transfer authorities.

First, the bill would authorizes the DCI to transfer funds after they have been appropriated "in coordination with" the Director of the Office of Management and Budget [H.R. 4584, see Title III Sec. 302(d)(B)]. Under current statute, and the Feinstein/Graham/Daschle/Harman bills, the DCI is required to seek the approval of the OMB Director before he can transfer funds [1947 National Security Act, see Sec. 104. [50 U.S.C. 403-4](d)].

The Goss bill also would remove from current statute a limitation that prohibits the DCI from transferring personnel within the NFIP for more than a year [H.R. 4584, see Title III, Sec.302(d)(F)]. The Feinstein/Graham/Daschle/Harman bills retain the one-year limitation.

The Goss bill also would allow an agency head to object to a DCI proposal to transfer funds or personnel, but would require an objecting agency head to provide the congressional intelligence committees written notice of their objection. The Goss language further requires that the DCI include within any fund or personnel transfer notification made to the congressional intelligence committees a copy of the written objection [H.R. 4584, see Title, III Sec.302(d)(B)]. Under current statute, the DCI cannot transfer of funds or personnel if the agency head objects in writing to the DCI [1947 National Security Act, see Sec.104 [50 U.S.C. 403-4] (d)(2)(A)(v) and (B)(iii)].

The Feinstein/Graham/Daschle bills, in eliminating any avenue for objection, would give the DNI unilateral fund and personnel transfer authority. The Harman bill would maintain an objection procedure that would require the objecting agency head to state his objection in writing to the President, who would ultimately arbitrate the disagreement.

Goss Would Strengthen DCI Personnel Authorities

The Goss bill would accord the DCI personnel termination authority contained in current statute, which grants hiring and firing authority over CIA employees [1947 National Security Act, see Sec.104[50U.S.C.403-4](h)]. The Goss bill also would add to the list of positions requiring DCI recommendation concurrence [in Goss, see Title III,Sec. 304]. Current statute requires that the Secretary of Defense obtain the DCI's concurrence in recommending to the President individuals for appointment to the positions of directors of NSA, NRO, and NGA. The Goss legislation would add to that list the positions of Director of the Defense Intelligence Ageny; the Assistant Secretary of State for Intelligence and Research; the Director of the Office of Intelligence of the Department of Energy; the Director of the Office of Counterintelligence of the National Nuclear Security Administration; the Assistant Secretary for Homeland Security for Information Analysis; and the Assistant Secretary for Intelligence and Analysis of the Department of Treasury.

Although required to seek the DCI's concurrence, the Goss bill would not change current statute that permits the respective agency heads from overriding the DCI's nonconcurrence.

With regard to the appointment of the Executive Assistant Director of the FBI's National Security Division, the Goss bill stipulates that the FBI Director and DCI jointly recommend to the Attorney General (AG) a candidate. The AG can ignore the recommendation. In contrast, current statute requires that the DCI be provided "timely notice" by the FBI Director before a recommendation is made to the AG.

Goss Would Eliminate Restriction of DCI Control over Internal Security Functions

The Goss bill would repeal the portion of the current statute that prohibits the DCI from exercising internal security functions [in Goss, see Title I, Sec.102(c)(1)]. The current prohibition against exercising policy, subpoena, or law enforcement powers would remain in place [1947 National Security Act, Sec.103[50 U.S.C.403- 1](d)(1). The Feinstein/Graham/Daschle/Harman bills would not lift the current restriction against exercising internal security functions.

ARGUMENTS OFFERED IN FAVOR OF ESTABLISHING A DNI

Supporters of the DNI concept argue that the DCI, who manages the IC and the CIA, and serves as the principal intelligence advisor to the President, has too many jobs, and that a DNI, unburdened by the need to manage the CIA, must be established if the IC is to be effectively managed.[17]

They also argue that a DNI must be empowered with two authorities the DCI now lacks: the authority to hire and remove IC agency heads in consultation with the Defense Secretary (currently, the Secretary of Defense selects principal IC agency heads for the three combat support agencies — NSA, NRO, and NGA — with the concurrence of the DCI); and the authority to move funding and personnel within or across IC agencies at any time during the year of execution with congressional approval.[18] Some proponents add a what they characterize as a third essential authority — the power to set standards for the IC's information infrastructure and personnel.[19]

From the prospective of proponents, failure to establish an empowered DNI with hiring, firing and budget authority will leave the IC with divided management of intelligence capabilities; lack of common standards and practices across the foreign-domestic intelligence divide; structural barriers that undermine the performance of joint intelligence work; and a weak capacity to set priorities and move resources.

They cite DCI George Tenet's 1998 "declaration of war"[20] on Osama Bin Laden and the corresponding lack of an integrated IC response as a clear indication of the need for a DNI.[21] Senator Bob Graham, a co-chairman of the 9/11 congressional joint inquiry, stated, "The intelligence community needs a leader with the clout to set common goals, establish priorities, knock heads and ensure that the American people are protected."[22] To accomplish that goal, DNI supporters argue, requires an empowered DNI with clear statutory end-to-end IC budget and personnel authorities, including authority over those large portions of the NFIP budget now controlled by the Secretary of Defense.

ARGUMENTS OFFERED IN OPPOSITION TO ESTABLISHING A DNI

Some opponents counter that although perhaps a good idea, establishing the position of DNI will have only a marginal impact, and assert that had this change been made prior to the September 11 attacks, it would not have significantly altered the way the U.S. dealt with Al Qaeda, and certainly would not have prevented the 9/11 attacks. They suggest that a more important step would be to hire more capable people throughout the IC.[23]

Other opponents contend that rather than strengthening control over the IC, the establishment of an NDI would actually weaken IC management. They assert a DNI would lose day-to-day control over the CIA, a natural power base. Without it, the DNI will lose influence, according to opponents. Admiral Bobby Inman, Former Deputy DCI and NSA director, said that DCIs rely on the CIA for their effectiveness and that a DNI "would be like the Drug Czar,"[24] a position that critics have argued has little management control over U.S. government agencies engaged in counternarcotics.

Other skeptics assert that establishing the position of DNI will only add another layer of bureaucracy, and risks disruption at a time when terrorists continue to threaten to attack the United States.

Critics are likely to also assert that centralized management control will be further weakened if the DNI is not granted meaningful hiring, firing and budget authority. They may argue that the DCI's relatively robust authorities to approve IC budgets and control budget reprogramming historically have gone largely unused in the face of DOD opposition. In other words, the DCIs have had the authority, but simply have chosen not to exercise it.

Some critics have voiced concern about the 9/11 Commission's recommendations to locate the office of the DNI in the executive office of the President. They contend that doing so risks the politicization of intelligence, would give the White House more direct control over covert operations, and would blur the line between foreign and domestic covert operations25. They also express concern that the Congress will experience greater difficulty in conducting oversight of the IC because the proximity of the DNI to the White House will more frequently raise the issue of executive privilege.

Some critics contend that the 9/11 Commissions concept of the NDI would shift too much influence over the IC to the Defense Department, because DOD would retain most of its roles under the commission's proposal, and indeed stands to gain influence. In contrast, other critics of the DNI concept oppose it because they believe DNI authorities could be so strengthened that DOD interests might suffer, if an empowered DNI were to favor providing more intelligence to policy makers rather than the warfighter. They argue the IC's three largest agencies — NSA, NRO and NGA — are combat support agencies that collect and disseminate intelligence affecting tactical military operations. It, therefore, is entirely appropriate, they argue, that the Secretary of Defense, rather than the DCI, control these agencies and the dollars that fund them, given that the needs of military commanders often differ from those of policymakers who generally are more interested in strategic intelligence.[26]

History of Recommendations to Centralize and Strengthen IC Leadership

The issue of centralized IC leadership was first addressed by the Second Hoover Commission in 1955. The following lists those Commissions, reports, individuals, executive orders and legislation that have addressed the issue of centralizing and strengthening IC leadership.[27]

Second Hoover Commission, 1955

The Commission on Organization of the Executive Branch of the Government, also known as the second Hoover Commission and chaired by former President Herbert Hoover, recommended that management of the CIA be turned over to an "executive officer," so that the DCI could focus attention on the IC.

The Schlesinger Report, 1971

President Nixon tasked the Office of Budget and Management to recommend changes in the IC's organization. Deputy OMB Director James R. Schlesinger, a future DCI, headed the effort and in his report considered the creation of a DNI, but in the end recommended that "a strong DCI who could bring intelligence costs under control and intelligence production to an adequate level of quality and responsiveness." Schlesinger criticized the IC's failure to coordinate resources, blaming the deficiency on the lack of a strong, central IC leadership that could "consider the relationship between cost and substantive output from a national perspective."

Murphy Commission, 1975

The Commission on the Organization of the Government for the Conduct of Foreign Policy, chaired by former Deputy Secretary of State Robert D. Murphy, noted that the DCI exercised direct control over the CIA but had only limited influence over the IC as a whole. But rather than recommending a structural change, the Commission said it was neither possible nor desirable to extend the DCI's control to the large part of the intelligence community that lies outside the CIA.

Church Committee, 1976

The Senate Select Committee to Study Governmental Operations with Respect to Intelligence Activities, known as the Church Committee and headed by Senator Frank Church, did not recommend establishing a DNI but urged that DCI authorities be strengthened by appropriating intelligence dollars directly to the DCI and by defining in statute DCI reprogramming authorities. The Committee also recommended that consideration be given to enhancing the DCI's management of the IC by relieving him of day-to-day management of the CIA.

Pike Committee, 1976

The House Select Committee on Intelligence, chaired by Congressman Otis G. Pike, recommended that the DCI should manage the IC as a whole and not exclusively the CIA. The Commission said the DCI should receive budget proposals from intelligence agencies comprising the community but did not indicate whether the DCI should have budget authority.

Clifford/Cline Proposals, 1976

Clark Clifford, a former Secretary of Defense under President Lyndon B. Johnson who had earlier participated in drafting legislation establishing the CIA, recommended that a new position of Director of General Intelligence be established and that a separate CIA director be responsible for managing the CIA.

Ray Cline, a former Deputy Director of the CIA, recommended that the DCI be given cabinet rank and broad supervisory authorities over the IC.

Charter Legislation, 1978

Following the establishment of the intelligence oversight committees in the Senate (1976) and in the House (1977), Congress considered charter legislation that, among other things, would have created a DNI to manage the IC. A presidentially selected deputy would manage CIA. In the face of strong opposition to the overall legislation, which also included language governing covert actions, the Committees did not report the respective bills.

Executive Branch Orders, 1976-1981

In an effort to head off further congressional action, President Gerald Ford in 1976 issued Executive Order (E.O.) 11905 naming the DCI as the President's primary intelligence advisor responsible for developing the NFIP.

President Jimmy Carter in 1978 issued E.O. 12036 (superseding E.O. 11905) more clearly defining the DCI's community-wide authority in areas relating to the budget, tasking, intelligence review, coordination, intelligence dissemination and foreign liaison.

President Ronald Reagan in 1981 continued the expansion of the DCI's community responsibilities and authorities, issuing E.O. 12333 (superseding E.O. 12036), which detailed the roles, responsibilities, missions, and activities of the IC. Executive Order 12333, which remains in effect today, granted the DCI more explicit authority over the development, implementation, and evaluation of the NFIP.

Turner Proposal, 1985

Admiral Stansfield Turner, former DCI under President Carter, recommended establishing a DNI to oversee the IC, and leaving responsibility for CIA day-to-day operations to a separate director of CIA.

Boren-Mccurdy, 1992

Senator David Boren and Congressman David McCurdy, respective chairmen of the Senate Select Committee on Intelligence (SSCI) and House Permanent Select on Intelligence (HPSCI), introduced legislation creating the position of a DNI with authority to program and reprogram funds, and creating a separate director of CIA. Boren and McCurdy failed to win adoption of their legislation in the face of opposition from DOD and the congressional Armed Services Committees.

Aspin-Brown Commission, 1996

The Commission on the Roles and Capabilities of the United States Intelligence Community (known as the Aspin-Brown Commission, after its respective chairmen Les Aspin and Harold Brown) concluded that the relationship between the DCI and Secretary of Defense should not be altered, but that the DCI should be given more time to manage the IC. The Commission recommended the creation of two deputies, one to help manage the IC and the other to manage the CIA.[28]

Specter/Combest, 1996

In the wake of the Aspin-Brown Commission report, Senator Arlen Specter and Congressman Larry Combest, respective chairmen of the SSCI and the HPSCI, sought to increase the clout of the DCI by giving him more control over the appointments of chiefs of defense-related agencies and the budgets of those agencies. But faced with intense opposition from the Pentagon and its congressional allies, they settled for more modest reform, agreeing to establish a new position of deputy DCI for community management and three assistant directors to oversee collection, analysis, and administration.[29] Although each were to be

Senate confirmed, only the deputy intelligence director for community management and the assistant director for administration have been confirmed. Neither the Clinton nor George W. Bush Administration has chosen to submit to the Senate for confirmation the names of individuals now serving as assistant directors for collection and analysis and production respectively.

Scowcroft Commission, 2001

A presidential commission chaired by retired Lt. Gen. Brent Scowcroft, the Chairman of President George W. Bush's Foreign Intelligence Advisory Board, reportedly recommended that the Pentagon should cede to the DCI control over DOD's three largest intelligence operations — NSA, NRO, and NGA. Although never made public, the report, according to media reports, was strongly opposed by Secretary of Defense Donald Rumsfeld.[30] The report never was formally presented to the President.

The Joint Inquiry into September 11 Terrorist Attacks

The Congressional Joint Inquiry Into Intelligence Community Activities Before and After the Terrorist Attacks of September 11, 2001, recommended that a new cabinet level Senate-confirmed DNI position be established and that a separate director be named to manage the CIA. The Joint Inquiry further recommended that the DNI be granted full IC budget execution and personnel programming authorities.[31]

The 9/11 Commission

The 9/11 Commission, in a report issued in July, 2004, recommended the establishment of a presidentially appointed, Senate-confirmed National Intelligence Director who would oversee national intelligence centers on specific subjects of interest across the U.S. government, manage the national intelligence program, oversee the agencies that contribute to it, and have hiring, firing and budgetary authority over the IC's 15 agencies. The Commission recommended that the director be located in the Executive Office of the President and that a deputy NID be established to oversee the day-to-day operations of the Central Intelligence Agency (CIA).

References

[1] See Richard A. Best, Jr. and H. Andrew Boerstling, "Proposals for Intelligence Reorganization, 1949-1996," in *IC21: Intelligence Community in the 21st Century*, Staff Study, House Permanent Select Committee on Intelligence, 1996. (The July, 2004 update of the CRS report is CRS Report RL32500, *Proposals for Intelligence Reorganization, 1949-2004*, by Richard A. Best.)

[2] The 911 Commission, formally known as the National Commission on the Terrorist Attacks Upon the United States, was created by congressional legislation and the signature of President George W. Bush in late 2002 (P.L. 107-306, Nov. 27, 2003). It was chartered to prepare a full and complete account of the circumstances surrounding the September 11, 2001 terrorist attacks, including preparedness for and the immediate response to the attacks. The Commission was also mandated to provide recommendations designed to guard against future attacks.

[3] The 9/11 Commission recommendations refer to a National Intelligence Director, or NID. Various congressional proposals name the position Director of National Intelligence, or DNI. This paper will use the "DNI" nomenclature.

[4] See The 9/11 Commission Report, National Commission on the Terrorist Attacks Upon the United States, July 22, 2004, p. 403.

[5] The Senate and House congressional oversight committees in 2002 initiated a joint inquiry into the September 11th, 2001 terrorist attacks on the United States. The inquiry issued its recommendations in December 2002, and an unclassified report of its findings in July, 2003.

[6] See U.S. Senate Select Committee on Intelligence and U.S. House Permanent Select Committee on Intelligence, *Joint Inquiry Into Intelligence Community Activities Before and After The Terrorist Attacks of September 11, 2001* (H.Rept. 107-792 and S.Rept. 107-351), errata print accompanying report, Recommendation No. 1, pp. 2-3.

[7] The National Foreign Intelligence Program is an aggregation of the budgets of the 15 agencies, including the CIA, which comprise the IC.

[8] See Report of the Commission on the Roles and Capabilities of the United States Intelligence Community, *Preparing for the 21st Century An Appraisal of the U.S. Intelligence*, March 1, 1996, P. xix.

[9] The National Security Agency is responsible for electronic intercepts; the National Reconnaissance Office designs, builds and operates the nation's reconnaissance satellites; and the National Geospatial-Intelligence Agency (NGA) provides geospatial intelligence, i.e. mapping. The NGA was formerly known as the National Image and Mapping Agency.

[10] See Studies in Intelligence, *The Need to Reorganize the Intelligence Community*, by Larry C. Kindsvater, Vol. 47, No. 1, 2003, P. 34.

[11] See Report of the Commission on the Roles and Capabilities of the United States Intelligence Community, *Preparing for the 21st Century An Appraisal of the U.S. Intelligence*, March 1, 1996, P. xix.

[12] Titled the "9/11 Memorial Intelligence Reform Act," the Graham legislation addresses seventeen of the Joint Inquiry's nineteen recommendations, and includes language creating a new Senate-confirmed position of Director of National Intelligence (DNI). See U.S. Senate Select Committee on Intelligence and U.S. House Permanent Select Committee on Intelligence, *Joint Inquiry Into Intelligence Community Activities Before and After The Terrorist Attacks of September 11, 2001* (H.Rept. 107-792 and S.Rept. 107-351), errata print accompanying report, Recommendation No. 1, pp. 2-3.

[13] Throughout this chapter, a distinction is drawn between budget-related *responsibilities* and budget-related *authorities*. These distinctions reflect identical distinctions made in the 1947 National Security Act.

[14] The concept of the year of budget execution is generally understood to mean the fiscal year in which monies already appropriated are actually spent

[15] The Joint Military Intelligence Program (JMIP) is composed of programs within the Defense Department that transcend the bounds of any one military service. JMIP historically has received over 10% of the intelligence budget.

[16] Tactical Intelligence and Related Activities (TIARA) is made up of the four service intelligence programs and intelligence for the Special Operations Command. TIARA historically has received about one-third of the intelligence budget.

[17] See the National Commission on Terrorist Attacks Upon the United States, *The 9/11Commission Report*, July, 2004, p. 409.

[18] See Studies in Intelligence, *The Need to Reorganize the Intelligence Community*, by Larry C. Kindsvater, Vol. 47, No. 1, 2003, Ibid. P. 36. With regard to the selection of agency heads, Sec. 106 (a) of the National Security Act [50 U.S.C. 403-6 (a) stipulates that if the DCI does not concur in the personnel recommendation by the Secretary of Defense , the Defense Secretary still may present his recommendation to the President without the DCI's concurrence, but include in the recommendation that the DCI does not agree with the recommendation.

[19] See the National Commission on Terrorist Attacks Upon the United States, *The 9/11Commission Report*, July 2004, p. 410.

[20] In December 1998, following the August bombings earlier that year of two American embassies in East Africa, Director of Central Intelligence George Tenet gave the following direction to his deputies: "We must now enter a new phase in our effort against Bin Ladin...We are at war..I want no resources or people spared in this effort, either inside the CIA or the Community." See Congressional Joint Inquiry into Intelligence Community Activities Before and After the Terrorist Attacks of September 11, 2001, pp 5-6.

[21] See U.S. Senate Select Committee on Intelligence and U.S. House Permanent Select Committee on Intelligence, *Joint Inquiry Into Intelligence Community Activities Before and After The Terrorist Attacks of September 11, 2001* (H.Rept. 107-792 and S.Rept. 107-351), errata print accompanying report, Recommendation No. 1, pp. 230-234.

[22] See Davies, Frank, "Graham, 2 allies introduce overhaul of U.S. intelligence," *The Miami Herald*, [http://www.Herald.com], August 1, 2003.

[23] See Richard A. Clarke, "Now For the Punches the 9/11 Commission Pulled," *Houston Chronicle*, July 28, 2004.

[24] See Staff Study, Permanent Select Committee on Intelligence, *IC21: Intelligence Community in the 21st Century*, 1996, P. 335, which includes a CRS report, *Proposals for Intelligence Reorganization, 1949-1996*, by Richard A. Best and Herbert Andrew Boerstling. See page 35 of the Best/Boerstling CRS report. (The update of the report is CRS Report RL32500, *Proposals for Intelligence Reorganization, 1949-2004*, by Richard A. Best.)

[25] See Walter Pincus, "9/11 Panel's Plan Would Reduce Influence of CIA," *Washington Post*, July 29, 2004, p. 6.

[26] See Pincus, Walter "Intelligence Shakeup Would Boost CIA; Panel Urges Transfer of NSA, Satellites, Imagery From Pentagon," *Washington Post*, November 8, 2001, P. A-1.

[27] For a more comprehensive treatment of IC reforms 1949-1996, see Staff Study, Permanent Select Committee on Intelligence, *IC21: Intelligence Community in the 21st Century*, 1996, P. 335, which included a CRS report, *Proposals for Intelligence Reorganization, 1949-1996*, by Richard A. Best and Herbert Andrew Boerstling. (An update to the CRS report is available in CRS Report RL32500, *Proposals for Intelligence Reorganization, 1949-2004*, by Richard A. Best.)

[28] See Report of the Commission on the Roles and Capabilities of the United States Intelligence Community, *Preparing for the 21st Century An Appraisal of the U.S. Intelligence*, March 1, 1996, pp. xix-xx.

[29] See *Congressional Quarterly Weekly*, "Defense and Intelligence Policy: Intelligence Reorganization," November 2, 1996.

[30] See Pincus, Walter "Intelligence Shakeup Would Boost CIA; Panel Urges Transfer of NSA, Satellites, Imagery From Pentagon", *Washington Post*, November 8, 2001, P. A-1.

[31] See U.S. Senate Select Committee on Intelligence and U.S. House Permanent Select Committee on Intelligence, *Joint Inquiry Into Intelligence Community Activities Before and After The Terrorist Attacks of September 11, 2001* (H.Rept. 107-792 and S.Rept. 107-351), errata print accompanying report, Recommendation No. 1, pp. 2-3.

APPENDIX 1. DNI LEGISLATION COMPARED TO CURRENT LAW

Table 1. Side-By-Side Comparison of DNI Legislation (H.R. 4104, S. 190, S. 1520, S. 6) and Current Statutes

H.R. 4104 (Harman)	S. 190 (Feinstein) S. 1520 (Graham), S. 6 (Daschle) [With one exception: the 3 bills contain identical DNI language.][a]	Current Law
Office of the Director of National Intelligence		
Title I, Sec. 101 would establish presidentially-nominated, Senate-confirmed positions of Director of National Intelligence (DNI) and Deputy DNI.	Comparable provision. See Sec. 2 (a) Sec. 102.	No comparable provision.
Title I, Sec. 101 would establish the Presidentially-nominated, Senate-confirmed position of Deputy DNI, who also would serve as Undersecretary of Defense for Intelligence.	No comparable provision.	No comparable provision.

Appendix 1 (Continued).

H.R. 4104 (Harman)	S. 190 (Feinstein) S. 1520 (Graham), S. 6 (Daschle) [With one exception: the 3 bills contain identical DNI language.][a]	Current Law
Title I, Sec. 101 would establish two presidentially-nominated, Senate-confirmed positions under the DNI — the Deputy DNI for Operations, and the Deputy DNI for Resources.	No comparable provision. Rather, Sec. 102 places under the DNI the currently existing presidentially-nominated, Senate-confirmed positions of the Deputy Director of Central Intelligence for Community Management, and the Assistant Directors for Collection and Analysis and Production, respectively, and re-titles each position.	The presidentially-nominated, Senate-confirmed positions of the Deputy Director of Central Intelligence for Community Management, and the Assistant Directors for Collection, and for Analysis and Production, respectively, are under the DCI.
Senior Advisor to the DNI for Homeland Security		
Title I, Sec. 101 stipulates that the DNI shall appoint a Senior Advisor to the DNI for Homeland Security.	No comparable provision.	No comparable provision.
General Counsel and Inspector General		
Title I, Sec. 101 would establish under the DNI Presidentially-nominated, Senate-confirmed positions of General Counsel and Inspector General.	Comparable provision. See Sec. 2 (a) Sec. 102 (a) (2), (H) and (I)	No comparable provision.
Alternative National Intelligence Council		
Title I, Sec. 101 would establish a National Intelligence Council Alternative Analysis Unit to review each national intelligence estimate produced by the National Intelligence Council.	No comparable provision.	No comparable provision.
DNI Responsibilities		
Title I, Sec. 101 stipulates that the DNI would "develop" an annual intelligence budget, and participate in the development with the Secretary of Defense of the annual budgets for the Joint Military Intelligence Program (JMIP) and the Tactical Intelligence and Related Activities Program (TIARA).	Comparable provision. See Sec. 2 (a) Sec. 102 (b).	No comparable provision.

Appendix 1 (Continued).

H.R. 4104 (Harman)	S. 190 (Feinstein) S. 1520 (Graham), S. 6 (Daschle) [With one exception: the 3 bills contain identical DNI language.][a]	Current Law
No comparable provision	The DNI shall develop an annual intelligence budget by "if necessary...directing the transfer of funds or personnel between elements of the Intelligence Community;" See Sec. 2 (a) Sec. 102 (b) (1) (c).	No comparable provision.
Title I, Sec. 101 stipulates that the DNI, among other tasks, would establish requirements and procedures for classifying and disseminating classified information; maximizing dissemination of classified information; ensuring that intelligence is portrayed accurately to the public; and, establishing unified procedures for granting access to sensitive compartmented information.	No comparable provision.	No comparable provision.
Comparable provision. See Title IV, Sec. 401.	Sec. 2(a) Sec. 103A(f) stipulates that the DNI, among other tasks, would rotate Intelligence Community (IC) personnel and make such rotation a factor to be considered for promotion to senior positions. This section also stipulates that the DNI cut costs by consolidating personnel, administrative and security programs.	No comparable provision.
DNI Authorities Approval of Budgets		
Title I, Sec. 103 (b) stipulates that the DNI would approve IC budgets.	Comparable provision (a) Sec. 103A (b). See Sec. 2	Comparable provision. See Sec. 104 (b) of the National Security Act of 1947.
Reprogramming		
Title I, Sec 103 (c) stipulates that no National Foreign Intelligence Program (NFIP) funds may be reprogrammed without the DNI's prior approval.	Comparable provision. See Sec. 2 (a) Sec. 103A (c).	Comparable provision. See Sec. 104 (c) of the National Security Act of 1947.

Appendix 1 (Continued).

H.R. 4104 (Harman)	S. 190 (Feinstein) S. 1520 (Graham), S. 6 (Daschle) [With one exception: the 3 bills contain identical DNI language.][a]	Current Law
Transfer of Funds or Personnel		
Title I, Sec. 103 (d) stipulates that the DNI may transfer, in accordance with procedures to be developed solely by the DNI, appropriated funds and authorized personnel for periods for up to a year.	Comparable provision. See Sec. 103A (d).	Comparable provision. See Sec. 104 (d) of the National Security Act of 1947, except the DCI, with input from agency heads, must develop transfer procedures.
Any Secretary, or head of department affected by such transfers, may object, but only in writing to the President.	No comparable provision.	No comparable provision.
Personnel Diversity		
Title IV, Sec. 401 stipulates that the DNI should ensure that there is personnel diversity within the IC.	No comparable provision.	No comparable provision.
Director of Central Intelligence		
Title I, Sec. 101 establishes a presidentially-nominated, Senate-confirmed position of Director of Central Intelligence (DCI), and prohibits the DCI from serving simultaneously as DNI.	Comparable provision. See Sec. 104(A).	Comparable provision, minus prohibition of DCI serving simultaneously as DNI
No comparable provision.	Sec. 104(d)(2) stipulates that the DCI provide overall IC direction for human collection.	Comparable provision. See Sec. 103(d)(2) of the National Security Act of 1947.
DNI Should Be Member of the Cabinet		
No comparable provision.	Sec. 104(2)(b) stipulates a sense of Congress that the DNI should be a cabinet-level officer.	No comparable provision.
Title I, Sec.102 would establish a Joint Tasking Organization within the DNI's office which would task IC collection, analysis and dissemin-ation. The DNI would appoint tasking directors from the CIA, National Security Agency and the National Geospatial-Intelligence Agency.	No comparable provision.	No comparable provision.

Appendix 1 (Continued).

H.R. 4104 (Harman)	S. 190 (Feinstein) S. 1520 (Graham), S. 6 (Daschle) [With one exception: the 3 bills contain identical DNI language.][a]	Current Law
Joint Tasking Organization		
Assistant Director of National Intelligence For Defense		
Title I, Sec.102 would establish the position of Assistant Director of National Intelligence For Defense who is assigned to the DNI and would coordinate the DOD intelligence elements not part of the NFIP.	No comparable provision.	No comparable provision.
Weapons of Mass Destruction Proliferation Threat Integration Center		
Title I, Sec. 103 would establish the Weapons of Mass Destruction Proliferation Threat Integration Center to provide integrated collection and analysis tasking with respect to weapons of mass destruction.	No comparable provision.	No comparable provision.
Establishment of Joint Intelligence Comptroller		
Title I, Sec. 104 would establish a Joint Intelligence Comptroller appointed jointly by the Secretary of Defense and the DNI. The Comptroller would report to the Undersecretary of Defense (comptroller) and provide both the Secretary of Defense and the DNI NFIP financial information insofar as that information would relate to elements of the.	No comparable provision.	No comparable provision.
IC under the jurisdiction of the Defense Department.		
Procedures For Use of Databases		
Title II, Sec. 201 stipulates that the President, acting jointly through the Attorney General and the DNI, would establish and implement intelligence sharing procedures.	No comparable provision.	No comparable provision.

Appendix 1 (Continued).

H.R. 4104 (Harman)	S. 190 (Feinstein) S. 1520 (Graham), S. 6 (Daschle) [With one exception: the 3 bills contain identical DNI language.][a]	Current Law
All-Source Analysis Training		
Title II, Sec. 202 stipulates that the DNI, acting through the DCI, would establish all-source analysis training programs for IC analysts.	No comparable provision.	No comparable provision.
Establishment of an Integrated Intelligence Network		
Title II, Sec. 203 stipulates that the DNI and the Secretary of Defense would develop an integrated IC communications network.	No comparable provision..	No comparable provision.
Acquisition Programs		
Title III, Sec. 301stipulates the establishment of a Joint Acquisition Office, headed by a director jointly appointed by the DNI and the Secretary of Defense. Would establish a senior acquisition executive within each of the IC agencies.	No comparable provision	No comparable provision.
Community Management		
Title IV, Sec. 402 stipulates that the DNI would establish policies and procedures for managing IC personnel, to be identified as intelligence community specialists.	No comparable provision.	No comparable provision.
Requirements For Service in Multiple Elements of the IC		
Title IV, Sec. 402 stipulates that no individual could be ap- pointed to a senior intelligence management position unless the individual had successfully completed a detail or assign- ment in more than two positions in elements of the IC outside the "home element" of the individual.	No comparable provision.	No comparable provision.
DNI Review of Promotion Lists		
Title IV, Sec. 402 stipulates that the DNI shall review recommendations of promotion boards for IC specialists.	No comparable provision.	No comparable provision.
Ten-Year Limit for DCI		
Title IV, Sec. 403 stipulates a ten-year term for the DCI.	No comparable provision.	No comparable provision.
Personnel Authorities		
Does not provide the DNI hiring and firing authority over IC agencies, but Title XI, Sec.404 strengthens some DNI authorities over IC appointments.	Adheres to current statute by not providing DNI hiring and firing authority over IC agencies; adheres to current statute by not strengthening DNI authorities over those currently enjoyed by the DCI.	Comparable provisions.

a. S. 6 would establish a ten-year term of service for the DNI. This side-by-side display describes only those bills that would establish the position of DNI.

In: Computer Networks, Policies and Applications
Editors: C. M. Berger and A. D. Streusse

ISBN: 978-1-61209-088-7

Chapter 7

GENERAL MANAGEMENT LAWS AND THE 9/11 COMMISSION'S PROPOSED OFFICE OF NATIONAL INTELLIGENCE DIRECTOR (NID) AND NATIONAL COUNTERTERRORISM CENTER (NCTC)*

Clinton T. Brass and Curtis W. Copeland

ABSTRACT

To improve the organization and capabilities of the U.S. intelligence community, the National Commission on Terrorist Attacks Upon the United States (9/11 Commission) recommended, among other things, that two entities be established — a National Intelligence Director (NID) and a National Counterterrorism Center (NCTC). The commission said that the NID should be located within the Executive Office of the President (EOP), and said the NCTC should report to the NID. Legislation designed to create these entities would place them either within the EOP (e.g., H.R. 5024 and H.R. 5050) or within an independent entity outside of the EOP (e.g., S. 2774 and H.R. 5040).

Questions have been raised regarding a number of issues associated with the establishment of the NID and the NCTC. One set of such questions centers on whether these proposed entities would be covered by various "general management laws" — broad statutes designed to regulate the activities, procedures, and administration of all or most executive branch agencies in such areas as regulatory and information management, financial management, procurement, and strategic planning. Earlier in 2004, CRS published a compendium of these general management laws and a companion report on major themes from these laws, with possible management policy options (CRS Report RL30795, *General Management Laws: A Compendium*; and CRS Report RL32388, *General Management Laws: Major Themes and Management Policy Options*).

This chapter examines whether more than 50 general management laws (e.g., the Administrative Procedure Act, the Inspector General Act, and the Competition in Contracting Act) would cover the NID and the NCTC under two scenarios: (1) if the entities were established within the EOP, and (2) if the entities were established as or

* This is an edited, reformatted and augmented edition of a United States Congressional Research Service publication, Report RL32592, dated September 21, 2004.

within independent entities in the executive branch. The analysis focuses on the *coverage* of the selected management laws, not on whether the laws are likely to *affect* or *apply* to the NID or the NCTC. Also, some of the laws allow exemptions or contain other provisions that can affect their coverage.

In general, the report indicates that more of these management laws would appear to cover the NID and NCTC if they were created as independent entities in the executive branch than if they were placed within the EOP. In many cases, it was unclear whether the statutes' definitions of covered "agencies" included entities within the EOP. The report also indicates that Congress, if it so desires, could indicate that the NID and the NCTC would be covered by these management laws by either (1) amending the text of each of the management laws or (2) delineating in the statute creating the entities the particular management laws that should cover them.

The National Commission on Terrorist Attacks Upon the United States (9/11 Commission), established by the Intelligence Authorization Act for Fiscal Year 2003 (P.L. 107-306; 116 Stat. 2383, at 2408), criticized the structure of the U.S. intelligence community.[1] To improve the intelligence community's organization and capabilities, the commission recommended (among other things) that two entities be established — a National Intelligence Director (NID) and a National Counterterrorism Center (NCTC). The commission said the NID should be located within the Executive Office of the President (EOP), and said the NCTC should report to the NID. Legislation designed to create these entities would place them either within the EOP or within an independent entity outside the EOP.

This chapter examines whether more than 50 general management laws (e.g., the Administrative Procedure Act, the Inspector General Act, and the Competition in Contracting Act) would cover the NID and the NCTC under two scenarios: (1) if the entities were established within the EOP, and (2) if the entities were established as or within independent entities in the executive branch. The report also discusses how Congress, if it so desires, can indicate that these proposed entities would be covered by these management laws. First, however, the report provides some background information on the 9/11 Commission's recommendations and recent efforts to restructure the intelligence community.

BACKGROUND

The 9/11 Commission questioned whether the U.S. government, and the intelligence community in particular, are organized adequately to direct resources and build the intelligence capabilities that the U.S. will need to counter terrorism, and to address the broader range of national security challenges in the coming decade. The commission recommended, among other things, that

> [t]he current position of Director of Central Intelligence should be replaced by a National Intelligence Director with two main areas of responsibility: (1) to oversee national intelligence centers on specific subjects of interest across the U.S. government and (2) to manage the national intelligence program and oversee the agencies that contribute to it.[2]

The commission said the NID would be the principal intelligence advisor to the President, and "would submit a unified budget for national intelligence that reflects priorities chosen by the National Security Council, an appropriate balance among the varieties of technical and human intelligence collection, and analysis."[3] The commission said the NID should be located in the EOP, would have a "relatively small staff of several hundred people," would be confirmed by the Senate, and would testify before Congress.[4] The commission also recommended

> the establishment of a National Counterterrorism Center (NCTC), built on the foundation of the existing Terrorist Threat Integration Center (TTIC). Breaking the older mold of national government organization, this NCTC should be a center for joint operational planning *and* joint intelligence, staffed by personnel from the various agencies. The head of the NCTC should have authority to evaluate the performance of the people assigned to the Center.[5]

The commission said the head of the NCTC should be appointed by the President, be confirmed by the Senate, testify before Congress, be equivalent in rank to a deputy head of a cabinet department, and report to the NID within the EOP.[6] The NCTC would override the current arrangement, in which the principal players — the Central Intelligence Agency and the Departments of Defense and Justice — were seen by the commission as operating without effective coordination and cooperation.

Legislation to Restructure the Intelligence Community

The 9/11 Commission's recommendations and several subsequent legislative initiatives are the most recent contributions to a debate on structural reform of the intelligence community that began nearly 50 years ago.[7] During the 107th and 108th Congresses, several Members of Congress have introduced legislation that would establish a position of Director of National Intelligence, or would strengthen the authority of the Director of Central Intelligence.[8]

For example, on September 7, 2004, a bipartisan group of Senators introduced legislation (S. 2774) to implement the recommendations made by the 9/11 Commission. In particular, the legislation would create an Office of the NID and the NCTC. However, in contrast to the commission's recommendation, the legislation would place the NID and the NCTC within an independent entity (the National Intelligence Authority), not within the EOP.[9] A House companion bill (H.R. 5040) was introduced on September 9, 2004, by Representative Christopher Shays (with 32 cosponsors).

On September 8, 2004, House Minority Leader Nancy Pelosi (with 117 cosponsors) introduced a bill (H.R. 5024) to implement the recommendations of the 9/11 Commission, including the establishment of the NID and NCTC. In contrast to the bills mentioned above (and consistent with the recommendations of the Commission), the legislation would place the proposed entities within the EOP. Another bill (H.R. 5050), introduced on September 9, 2004, by Representative Sheila Jackson-Lee, would establish a Director of National Intelligence within the EOP.

A news report stated that a "leadership bill" will be introduced by the end of September, reportedly by Speaker of the House J. Dennis Hastert.[10]

On September 15, 2004, Chairman Susan Collins and Ranking Member Joseph Lieberman of the Senate Committee on Governmental Affairs announced legislation to "reform the Executive Branch of the nation's intelligence community."[11] According to a fact sheet, the proposed entities would be placed in a National Intelligence Authority, not within the EOP.[12] Other legislation designed to establish the NID and the NCTC continues to be introduced.

On August 27, 2004, President George W. Bush issued Executive Order 13355, which was intended to strengthen the position of the Director of Central Intelligence.[13] On the same day, he also issued Executive Order 13354, which established the NCTC, over which the Director of Central Intelligence "shall have authority, direction, and control."[14] On September 8, 2004, President Bush indicated that the Administration sought legislation that would establish the NID position within the Executive Branch, but not within the EOP.

NID and NCTC and the Coverage of General Management Laws

General Management Laws

Questions have been raised regarding a number of issues associated with the establishment of the NID and the NCTC. One set of such questions centers on whether these proposed entities would be covered by various "general management laws" — broad statutes designed to regulate the activities, procedures, and administration of all or most executive branch agencies in such areas as regulatory and information management, financial management, procurement, and strategic planning. Earlier in 2004, CRS published a compendium of these general management laws and a companion report on major themes from these laws, with possible management policy options.[15]

To determine the coverage of these management laws in relation to the proposed NID and NCTC, CRS examined the scope of the laws under two scenarios — if the entities were established (1) within the EOP or (2) as or within independent entities in the executive branch. In both scenarios, CRS assumed that the entities would be established by Congress. This review included more than 50 such laws, but did not include laws related to human resource management (Title 5 of the *United States Code*) or intergovernmental relations (e.g., the Intergovernmental Cooperation Act).

Table Showing Coverage of Laws

Table Structure and Contents

A table indicating the coverage of each management law in each scenario is included at the end of this chapter. The first column of the table identifies the general management laws and their locations in the table of contents for the previously mentioned compendium of these laws. The next three columns provide CRS's best judgment regarding the coverage of each of the laws if the NID and the NCTC were established in the EOP — i.e., whether (1) the entities appear to be covered by the law; (2) the entities do not appear to be covered by the law; or (3) it is not clear if the entities would be covered. The next three columns indicate CRS's best judgments regarding the coverage of these laws if the NID and NCTC were established as independent entities within the executive branch. The last column indicates the CRS subject matter experts who may be contacted for further information about each of the general management laws. (For legal issues that relate to definitions of "agency," T. J. Halstead (7-7981) and Morton Rosenberg (7-7480), in CRS's American Law Division, are also available for further information and assistance.)

Summary of Findings

In general, this analysis indicated that more of these general management laws would appear to cover the NID and the NCTC if they were created as or within independent entities in the executive branch than if they were established within the EOP. In many cases, it was unclear whether the statutes' definitions of covered "agencies" included entities within the EOP. Therefore, if Congress chose to establish the proposed entities, the coverage of these management laws would appear to depend on where the entities were established.

Within the EOP

If the NID and the NCTC were established within the EOP, about half (26) of the general management laws included in this chapter would appear to cover, in whole or in part, the proposed entities.[16] Eleven of the laws would not appear to cover the entities within the EOP, and it was not clear whether they would be covered by 18 of the laws. For example, several of the statutes indicated they covered any "department, agency, or instrumentality" of the U.S. government, but it was not clear whether this construction would include the EOP or any of its constituent parts. Of the 11 laws that did not appear to cover the entities, four did not cover the entities because the laws had expired. (These laws were included in the CRS compendium because of continuing congressional interest in the subjects.) In other of these "not covered" cases, the laws (e.g., the Inspector General Act of 1978 and the Federal Managers' Financial Integrity Act of 1982) specifically list the covered agencies, and do not include the EOP; nor would they include the proposed entities.

Independent Entities

If the NID and the NCTC were established as or within independent entities in the executive branch, more than 75% (42) of the general management laws would appear to cover, in whole or in part, the proposed entities. In many cases, where it was unclear that a particular management law covered an "instrumentality" or "independent entity" within the EOP, it was more clear that the law would cover an independent agency or other entity within the executive branch. Of the 12 laws that did not appear to cover the entities, four would not do so because the laws have expired, and others did not appear to cover the entities because they were not included in a list of covered agencies. In only one case was it unclear whether the management law would apply to an independent entity.

Considerations Regarding Table Contents

Several points are important to consider with regard to the table's contents: (1) the analysis focuses on the *coverage* of the selected management laws with respect to the NID and the NCTC, not whether the laws are likely to *affect* or *apply to* these proposed entities; (2) the coverage of some of the laws in these scenarios is unclear; (3) Congress has previously exempted certain agencies from the coverage of some of the management laws; and (4) certain provisions in some of the laws may affect their coverage.

"Coverage" Versus "Application" of the Laws

This chapter focuses on whether an entity is *covered by* a statute, not on whether the law is likely to *affect* or *apply to* the entity's operations. For example, the Paperwork Reduction Act (PRA) specifically identifies the EOP in its definition of a covered "agency." Therefore, if the NID and the NCTC were created as part of the EOP, we concluded that they would be *covered by* the PRA. However, if these entities never imposed any paperwork requirements on the public that triggered the PRA's requirements, the act would not *affect* or *apply to* their operations. Similarly, the National Environmental Policy Act might *cover* these entities (particularly if they were established as independent entities within the executive branch), but they might never take actions that triggered the requirement in the act for an environmental impact statement.

Coverage Sometimes Unclear

CRS determined whether the selected management laws would appear to cover the NID and the NCTC in particular scenarios based on the definitions and other terms provided in these laws — most commonly, how the terms "agency" or "executive agency" were defined. In some cases, however, determining the coverage of these management laws was difficult. As noted previously, this difficulty was most common in determining whether certain statutes applied to entities within the EOP. For example, the Federal Records Act covers any "independent establishment" in the executive branch, but this term is not defined in the act. Given the nature and missions of the various offices within the EOP (i.e., to help develop and implement the policies and programs of the President), it is not clear whether the NID or the NCTC would be considered "independent establishments" for purposes of the act if established within the EOP. In some cases, the coverage of these management laws might turn on the interpretations of the courts.

It should also be noted that coverage of the NID and the NCTC would not be ensured even in instances where the statutory definition either seems broad enough to include the EOP, or specifically includes the EOP within its ambit. For example, the Administrative Procedure Act (APA) defines "agency" to include "each authority of the Government of the United States, whether or not it is within or subject to review by another agency."[17] While this definition could arguably apply to any governmental entity, the United States Court of Appeals for the District of Columbia has established that "the APA confers agency status on any administrative unit with substantial independent authority in the exercise of specific functions."[18] This standard has been of particular importance in the context of the Freedom of Information Act (FOIA), even though the act specifically includes the EOP in its definition of "agency."[19] In *Meyer v. Bush*, for instance, the United States Court of Appeals for the District of Columbia determined that the Task Force on Regulatory Relief was not subject to FOIA even though it was located in the EOP, based on the fact that its duties were advisory in nature and that it "was not a body with 'substantial independent authority' to direct executive branch officials."[20] Accordingly, the determination in the attached chart that the APA and FOIA would appear to cover the NID and the NCTC as part of the EOP is predicated upon the assumption that these entities would, in fact, be imbued with "substantial independent authority" in the exercise of their functions.[21]

Exemptions from Coverage

As the attached table indicates, many of the management laws that CRS examined would appear to cover the proposed NID and NCTC as part of the EOP, as independent entities, or in either situation. Congress could, however, exempt one or both of these entities from the coverage of certain laws — just as it has done for other entities engaged in sensitive or intelligence-related activities. For example, Congress specifically exempted the Central Intelligence Agency (CIA) from requirements of the Federal Advisory Committee Act (Section 4 of 5 U.S.C. Appendix 1), from "law and regulations relating to the expenditure of government funds" (50 U.S.C. § 403j(b)), and from the Government Performance and Results Act (GPRA; 5 U.S.C. § 306(f)).[22] Some of the management laws also allow other entities to exempt certain agencies or activities from coverage. For example, GPRA allows the Office of Management and Budget (OMB) to exempt agencies with annual outlays of $20 million or less from certain requirements.

Certain Provisions can Affect Coverage

For some of the general management laws included in the attached table, even though these laws would normally cover the proposed NID and NCTC, certain provisions could affect this coverage. For example, the entry for the Davis-Bacon Act notes that, while the act would appear to cover the NID and the NCTC (either as part of the EOP or as independent entities), the President can suspend the act during a "national emergency." Also, the Chief Financial Officers (CFO) Act of 1990 technically does not cover the EOP, but a 1999 amendment to the act created a separate CFO for the EOP. The President can, under this amendment, determine what authority and functions apply to the CFO for the EOP. Although the table notes several of these provisions, it should not be relied upon to highlight all important nuances in the coverage of general management laws.[23]

Congress Could Indicate whether Management Laws Cover NID and NCTC

If Congress decided to create the NID and the NCTC and wanted to indicate that these entities should be covered by certain management laws, two general approaches would be available — (1) amending the text of each of the relevant management laws to indicate that the entities being created should be considered covered by these laws, or (2) including a provision in the statute creating these entities indicating which of the management laws cover them. Although either of these general approaches would accomplish this objective, each has particular advantages and disadvantages.

Amending the Management Laws

In the first approach, Congress, after creating the NID and the NCTC, could consider amending the text of selected general management laws to indicate clearly that they cover the new entities. For example, the Chief Financial Officers (CFO) Act of 1990 lists the agencies covered by its requirements at 31 U.S.C. § 901(b). If Congress wanted to have the CFO Act cover one or both new entities, Congress could add the entity or entities to this list, and thereby put the complete list of covered agencies at one location in the *United States Code*. Similar action would be required for each of the statutes that Congress considers relevant. A possible advantage of this approach would be that, in comparison to the second alternative, it would be easier for an interested party to determine the coverage of a specific general management law when specific agencies were delineated. However, this approach also could have several disadvantages. For example, making changes in statutory language to dozens of particular statutes would likely be significantly more difficult than the alternative approach, and more likely to result in jurisdictional issues among congressional committees.

Identification of Management Laws in Statute Creating NID/NCTC

In the second approach, Congress could specifically delineate the particular management laws that should cover the NID and the NCTC in the statute creating them. This listing of management laws could include all such laws that Congress considers relevant (including those that already would appear to cover the proposed entities) in order to clearly establish congressional intent. Alternatively, the listing could focus only on either those statutes that do not appear to cover the new entities, or those for which coverage is uncertain.

In either case, the statute creating the entities could state that they "shall be considered to be covered by the requirements of ...," with a listing of all relevant management laws provided (e.g., the Federal Register Act and the APA). Or, more specifically, the statute could include language defining a covered "agency" (or other term, depending on the specific management laws) by referring to the definitions in the management laws themselves, such as the following:

> The [NID/NCTC] shall be considered to be a covered "agency" for purposes of the Administrative Procedure Act, as defined at 5 U.S.C. 551(1); ... [add other applicable management laws and code citations].

Some potential advantages of this general approach include relative ease of statutory drafting and fewer jurisdictional issues than with the first approach (amending each management law). However, this approach might make it more difficult for an interested party to determine the coverage of a given law (particularly for those laws that list the covered agencies), because to do so, the party would have to find all provisos of interest, which might be included in many parts of the *United States Code*. For example, if the statute creating the NID and the NCTC indicated that one or both of the entities should be covered by the CFO Act, an interested party attempting to compile a list of agencies with CFOs would have to know to examine not only the CFO Act, but also the statute creating the new entities.

Table 1. Coverage of General Management Laws and the 9/11 Commission's Proposals

General Management Law (location in CRS Report RL30795)	If the Office of the National Intelligence Director (NID) and National Counterterrorism Center (NCTC) were established ...						CRS Analyst, Telephone Number
	... in the Executive Office of the President (EOP), the NID and the NCTC appear to be:			... as independent entities, outside the EOP, but in the executive branch, the NID and the NCTC appear to be:			
	Covered	Not Covered	Not Clear	Covered	Not Covered	Not Clear	
Federal Register Act (I.A. in RL30795)	X (Falls within the definition of "agency" in the act (44 U.S.C. § 1501) for required or authorized publication of designated documents (44 U.S.C. § 1505).)			X (Falls within the definition of "agency" in the act (44 U.S.C. § 1501) for required or authorized publication of designated documents (44 U.S.C. § 1505).)			Harold C. Relyea, 7-8679
Administrative Procedure Act (APA) (I.B. in RL30795)	X (The definition of "agency" in the act (5 U.S.C. § 551(1)) covers each authority of the government, "whether or not it is within or subject to review by another agency." However, if either proposed entity were strictly advisory in nature, it would not be covered.)			X (The definition of "agency" in the act (5 U.S.C. § 551(1)) covers each authority of the government, "whether or not it is within or subject to review by another agency." However, if either proposed entity were strictly advisory in nature, it would not be covered.)			Morton Rosenberg, 7-7480; T. J. Halstead, 7-7981
Federal Records Act and Related Chapters of Title 44 (I.C. in RL30795)			X (The act covers any "executive agency," which is defined (40 U.S.C. § 472(a)) as any executive department or "independent establishment" in the executive branch. It appears that an entity within the EOP would not be considered an "independent establishment," but it is not clear.)	X (The act covers any "executive agency," which is defined (40 U.S.C. § 472(a)) as any executive department or "independent establishment" in the executive branch. An independent entity would likely be considered an "independent establishment.")			Harold C. Relyea, 7-8679

Table 1. (Continued).

General Management Law (location in CRS Report RL30795)	If the Office of the National Intelligence Director (NID) and National Counterterrorism Center (NCTC) were established ...						CRS Analyst, Telephone Number
	... in the Executive Office of the President (EOP), the NID and the NCTC appear to be:			... as independent entities, outside the EOP, but in the executive branch, the NID and the NCTC appear to be:			
	Covered	Not Covered	Not Clear	Covered	Not Covered	Not Clear	
Congressional Review of Regulations Act (CRA) (I.D. in RL30795)	X (The definition of "agency" in the act refers to the definition in the APA (5 U.S.C. § 551(1)), which covers each authority of the government, "whether or not it is within or subject to review by another agency." As with the APA, however, if either organization were strictly advisory in nature, it would not be covered.)			X (The definition of "agency" in the act refers to the definition in the APA (5 U.S.C. § 551(1)), which covers each authority of the government, "whether or not it is within or subject to review by another agency." As with the APA, however, if either organization were strictly advisory in nature, it would not be covered.)			Morton Rosenberg, 7-7480
Freedom of Information Act (I.E. in RL30795)	X (Falls within the definition of "agency" in the act (5 U.S.C. § 552(f)), which specifically includes the EOP. As with the APA, however, if either organization were strictly advisory in nature, it would not be covered.)			X (Falls within the definition of "agency" in the act (5 U.S.C. § 552(f)), which includes any "establishment" in the executive branch. As with the APA, however, if either organization were strictly advisory in nature, it would not be covered.)			Harold C. Relyea, 7-8679

Table 1. (Continued).

General Management Law (location in CRS Report RL30795)	If the Office of the National Intelligence Director (NID) and National Counterterrorism Center (NCTC) were established ...						CRS Analyst, Telephone Number
	... in the Executive Office of the President (EOP), the NID and the NCTC appear to be:			... as independent entities, outside the EOP, but in the executive branch, the NID and the NCTC appear to be:			
	Covered	Not Covered	Not Clear	Covered	Not Covered	Not Clear	
Privacy Act (I.F. in RL30795)	X (Falls within the definition of "agency" in the act (5 U.S.C. § 552a(a)(1), referencing 5 U.S.C. § 552(f)), which specifically includes the EOP. As with the APA, however, if either organization were strictly advisory in nature, it would not be covered.)		X (Falls within the definition of "agency" in the act (5 U.S.C. § 552a(a)(1), referencing 5 U.S.C. § 552(f)), which includes any "establishment" in the executive branch. As with the APA, however, if either organization were strictly advisory in nature, it would not be covered.)				Harold C. Relyea, 7-8679
Federal Advisory Committee Act (I.G. in RL30795)	X (Falls within the definition of "agency" in the act (5 U.S.C. § 551(1)), although Congress currently exempts advisory committees used by the CIA.)			X (Falls within the definition of "agency" in the act (5 U.S.C. § 551(1)), although Congress currently exempts advisory committees used by the CIA.)			Stephanie Smith, 7-8674
Government in the Sunshine Act (I.H. in RL30795)		X (Would not cover the proposed entities unless established as collegially headed executive branch agencies with two or more members, most of whom are appointed by the President and confirmed by the Senate.)			X (Would not cover the proposed entities unless established as collegially headed executive branch agencies with two or more members, most of whom are appointed by the President and confirmed by the Senate.)		Henry Hogue, 7-0642

Table 1. (Continued).

General Management Law (location in CRS Report RL30795)	If the Office of the National Intelligence Director (NID) and National Counterterrorism Center (NCTC) were established ...						CRS Analyst, Telephone Number
	... in the Executive Office of the President (EOP), the NID and the NCTC appear to be:			... as independent entities, outside the EOP, but in the executive branch, the NID and the NCTC appear to be:			
	Covered	Not Covered	Not Clear	Covered	Not Covered	Not Clear	
Paperwork Reduction Act (I.I. in RL30795)	X (Falls within the definition of "agency" in the act (44 U.S.C. § 3502), which specifically includes the EOP.)			X (Falls within the definition of "agency" in the act (44 U.S.C. § 3502), which includes virtually any "establishment" in the executive branch.)			Harold C. Relyea, 7-8679
Regulatory Flexibility Act of 1980 (I.J. in RL30795)	X (The definition of "agency" in the act refers to the definition in the APA (5 U.S.C. § 551(1)), which covers each authority of the government, "whether or not it is within or subject to review by another agency." As with the APA, however, if either organization were strictly advisory in nature, it would not be covered.)			X (The definition of "agency" in the act refers to the definition in the APA (5 U.S.C. § 551(1)), which covers each authority of the government, "whether or not it is within or subject to review by another agency." As with the APA, however, if either organization were strictly advisory in nature, it would not be covered.)			Curtis Copeland, 7-0632
Negotiated Rulemaking Act (I.K. in RL30795)	X (The definition of "agency" in the act (5 U.S.C. § 562(1)) refers to the definition in the APA (5 U.S.C. § 551(1)), which covers each authority of the government, "whether or not it is within or subject to review by another agency." As with the APA, however, if either organization were strictly advisory in nature, it would not be covered.)			X (The definition of "agency" in the act (5 U.S.C.§ 562(1)) refers to the definition in the APA (5 U.S.C. § 551(1)), which covers each authority of the government, "whether or not it is within or subject to review by another agency." As with the APA, however, if either organization were strictly advisory in nature, it would not be covered.)			Curtis Copeland, 7-0632

Table 1. (Continued).

General Management Law (location in CRS Report RL30795)	If the Office of the National Intelligence Director (NID) and National Counterterrorism Center (NCTC) were established ...						CRS Analyst, Telephone Number
	... in the Executive Office of the President (EOP), the NID and the NCTC appear to be:			... as independent entities, outside the EOP, but in the executive branch, the NID and the NCTC appear to be:			
	Covered	Not Covered	Not Clear	Covered	Not Covered	Not Clear	
National Environmental Policy Act (I.L. in RL30795)			X (The requirements in 42 U.S.C. § 4332 cover "all agencies of the Federal Government," which is not defined. It is not clear whether "all agencies" includes the EOP.)	X (The requirements in 42 U.S.C. § 4332 cover "all agencies of the Federal Government," and other independent agencies are covered by the act.)			Harry Steven Hughes, 7-726
E-Government Act (Titles II, III, V) (I.M. in RL30795)	X (Section 201 of the act says the definitions in the PRA generally apply, which defines "agency" (44 U.S.C. § 3502(1)) as including the EOP.)			X (Section 201 of the act says the definitions in the PRA generally apply, which defines "agency" (44 U.S.C. § 3502(1)) as including virtually any establishment in the executive branch.)			Harold C. Relyea, 7-8679; Jeffrey W. Seifert, 7-0781
Federal Information Security Management Act of 2002 (I.N. in RL30795)	X (The act amends the PRA, which specifically defines "agency" (44 U.S.C. § 3502) as including the EOP.)			X (The act amends the PRA, which defines "agency" (44 U.S.C. § 3502) to include virtually any establishment in the executive branch.)			John Moteff, 7-1435
Data Quality Act (Information Quality Act) (I.O. in RL30795)	X (The act amends the PRA, which specifically defines "agency" (44 U.S.C. § 3502) as including the EOP.)			X (The act amends the PRA, which defines "agency" (44 U.S.C. § 3502) to include virtually any establishment in the executive branch.)			Curtis Copeland, 7-0632; Michael Simpson, 7-7010

Table 1. (Continued).

General Management Law (location in CRS Report RL30795)	If the Office of the National Intelligence Director (NID) and National Counterterrorism Center (NCTC) were established ...						CRS Analyst, Telephone Number
	... in the Executive Office of the President (EOP), the NID and the NCTC appear to be:			... as independent entities, outside the EOP, but in the executive branch, the NID and the NCTC appear to be:			
	Covered	Not Covered	Not Clear	Covered	Not Covered	Not Clear	
Inspector General Act of 1978 (II.A. in RL30795)		X (The act (5 U.S.C. App., § 11) lists the specific departments and agencies that are covered; it does not include either the EOP and would not include the proposed entities. However, note 50 U.S.C. § 403q establishing an inspector general for the CIA.)			X (The act (5 U.S.C. App., § 11) lists the specific departments and agencies that are covered, and would not include the proposed entities. However, note 50 U.S.C. § 403q establishing an inspector general for the CIA.)		Frederick Kaiser, 7-8682
Government Performance and Results Act of 1993 (II.B. in RL30795)			X (GPRA defines "agency" (5 U.S.C. § 306(f), 31 U.S.C. § 1115(f), and 5 U.S.C. § 105) as "an Executive department, a Government corporation, and an independent establishment," but exempts the CIA due to its classified activities. It is not clear if this definition includes the EOP. OMB may exempt agencies with annual outlays of $20 million or less.)			X (GPRA defines "agency" (5 U.S.C. § 306(f), 31 U.S.C. § 1115(f), and 5 U.S.C. § 105) as "an Executive department, a Government corporation, and an independent establishment," but exempts the CIA due to its classified activities. P.L. 104-293, Sections 803 and 808, have separate reporting requirements for intelligence agencies. OMB may exempt agencies with annual outlays of $20 million or less.)	Genevieve J. Knezo, 7-6610

Table 1. (Continued).

General Management Law (location in CRS Report RL30795)	If the Office of the National Intelligence Director (NID) and National Counterterrorism Center (NCTC) were established ...						CRS Analyst, Telephone Number
	... in the Executive Office of the President (EOP), the NID and the NCTC appear to be:			... as independent entities, outside the EOP, but in the executive branch, the NID and the NCTC appear to be:			
	Covered	Not Covered	Not Clear	Covered	Not Covered	Not Clear	
Clinger-Cohen Act of 1996 (II.C. in RL30795)		X (Different provisions of the Clinger-Cohen Act (e.g., 40 U.S.C. 1425(c)) identify different sets of agencies as being covered (e.g., 5 U.S.C. § 101, 31 U.S.C. § 901(b)). However, these provisions do not indicate that EOP is covered.)	X (Provisions of the act (e.g., 40 U.S.C. § 1421 et seq.) that cover an "executive agency" (defined at 40 U.S.C. § 1401, which points to 41 U.S.C. § 403(1)) cover an "independent establishment." It is unclear if an entity within the EOP is considered an "independent establishment.")	X (Provisions of the act (e.g., 40 U.S.C. § 1421 et seq.) that cover an "independent establishment" (included in the definition of "executive agency" defined at 40 U.S.C. § 1401, which points to 41 U.S.C. § 403(1)) would appear to cover these independent entities.)	X (Provisions of the act that cover specified departments or agencies (e.g., CFO Act agencies as identified in 31 U.S.C. § 901(b)) do not apply to other independent entities in the executive branch.)		Jeffrey W. Seifert, 7-0781
Antideficiency Act (III.A. in RL30795)	X (The act covers any "officer or employee of the United States Government" (31 U.S.C. § 1341(a)(1)), which presumably includes the EOP.)			X (The act covers any "officer or employee of the United States Government" (31 U.S.C. § 1341(a)(1)), which presumably includes independent entities in the executive branch.)			Virginia McMurtry, 7-8678
Budget and Accounting Act of 1921 (III.B. in RL30795)	X (Unless otherwise provided for, the proposed entities' budgets would be covered by the act in the same way as other parts of the EOP.)			X (Unless otherwise provided for, the proposed entities' budgets would be covered by the act in the same way as other independent entities in the executive branch.)			James V. Saturno, 7-2381

Table 1. (Continued).

General Management Law (location in CRS Report RL30795)	If the Office of the National Intelligence Director (NID) and National Counterterrorism Center (NCTC) were established ...						CRS Analyst, Telephone Number
	... in the Executive Office of the President (EOP), the NID and the NCTC appear to be:			... as independent entities, outside the EOP, but in the executive branch, the NID and the NCTC appear to be:			
	Covered	Not Covered	Not Clear	Covered	Not Covered	Not Clear	
Budget and Accounting Procedures Act of 1950 (III.C. in RL30795)	X (The act is generally applicable to federal agencies. However, note 50 U.S.C. § 403j(b) exempting the CIA from laws and regulations relating to the expenditure of government funds.)			X (The act is generally applicable to federal agencies. However, note 50 U.S.C. § 403j(b) exempting the CIA from laws and regulations relating to the expenditure of government funds.)			Bob Lyke, 7-7355
Balanced Budget and Emergency Deficit Control Act (III.D. in RL30795)		X (This law expired at the end of FY2002.)			X (This law expired at the end of FY2002.)		Robert Keith, 7-8659
Budget Enforcement Acts of 1990 and 1997 (III.E. in RL30795)		X (These laws expired at the end of FY2002.)			X (These laws expired at the end of FY2002.)		Robert Keith, 7-8659; Bill Heniff, 7-8646
Congressional Budget and Impoundment Act (III.F. in RL30795)	X (The entities' budgets would be covered by the act in the same way as other parts of the EOP.)			X (The entities' budgets would be covered by the act in the same way as other independent entities in the executive branch.)			James V. Saturno, 7-2381
Chief Financial Officers Act of 1990 (III.G. in RL30795)	X (The original act listed (in 31 U.S.C. § 901(b)) specific covered departments and agencies, but did not include the EOP and would not include the proposed entities. However, the EOP has a statutory CFO (31 U.S.C. § 901(c)(1)-(4)), distinct from the original CFO Act agencies. The authority and functions of the EOP CFO are at the discretion of the President.)				X (The act lists (in 31 U.S.C. § 901(b)) the specific departments and agencies that are covered, and would not include these entities.)		Virginia McMurtry, 7-8678

Table 1. (Continued).

General Management Law (location in CRS Report RL30795)	If the Office of the National Intelligence Director (NID) and National Counterterrorism Center (NCTC) were established ...						CRS Analyst, Telephone Number
	... in the Executive Office of the President (EOP), the NID and the NCTC appear to be:			... as independent entities, outside the EOP, but in the executive branch, the NID and the NCTC appear to be:			
	Covered	Not Covered	Not Clear	Covered	Not Covered	Not Clear	
Government Management Reform Act of 1994 (III.H. in RL30795)	X (Certain provisions of the act (e.g., direct deposit of federal wage, salary, and retirement payments) cover all agencies, including the EOP.)	X (The financial reporting requirements in section 405 cover only agencies listed in 31 U.S.C. § 901(b). The EOP is not listed, nor would the proposed entities be listed.)		X (Certain provisions of the act (e.g., direct deposit of federal wage, salary, and retirement payments) cover all agencies, including the independent entities.)	X (The financial reporting requirements in section 405 cover only agencies listed in 31 U.S.C. § 901(b). The proposed entities would not be listed.)		Virginia McMurtry, 7-8678
Accountability of Tax Dollars Act of 2002 (III.I. in RL30795)			X (Definition of a "covered executive agency" (31 U.S.C. § 3515(f)) suggests that the act would cover the proposed entities in the EOP, but ultimately is unclear.	X (Definition of a "covered executive agency" (31 U.S.C. § 3515(f)) indicates that the act would cover the proposed entities.)			Virginia McMurtry, 7-8678
Federal Managers' Financial Integrity Act of 1982 (III.J. in RL30795)		X (The act refers to the definition of "agency" in 31 U.S.C. § 901(b), which lists specific agencies that are covered and does not include either the EOP or these proposed entities.)			X (The act refers to the definition of "agency" in 31 U.S.C. § 901(b), which lists specific agencies that are covered and would not include the proposed entities.)		Frederick Kaiser, 7-8682

Table 1. (Continued).

General Management Law (location in CRS Report RL30795)	If the Office of the National Intelligence Director (NID) and National Counterterrorism Center (NCTC) were established ...						CRS Analyst, Telephone Number
	... in the Executive Office of the President (EOP), the NID and the NCTC appear to be:			... as independent entities, outside the EOP, but in the executive branch, the NID and the NCTC appear to be:			
	Covered	Not Covered	Not Clear	Covered	Not Covered	Not Clear	
Federal Financial Management Improvement Act of 1996 (III.K. in RL30795)		X (This act covers only CFO Act agencies (see 31 U.S.C. § 901(b)), which do not include either the EOP or these proposed entities.)			X (This act covers only CFO Act agencies (see 31 U.S.C. § 901(b)), which would not include the proposed entities.)		Virginia McMurtry, 7-8678
Federal Credit Reform Act of 1990 (III.L. in RL30795)	X (The act (2 U.S.C. § 661a) covers any "disbursement of funds by the Government to a non-Federal borrower under a contract that requires repayment of such funds with or without interest." Therefore it appears to cover such activities within the EOP.)			X (The act (2 U.S.C. § 661a) covers any "disbursement of funds by the Government to a non-Federal borrower under a contract that requires repayment of such funds with or without interest." Therefore it appears to cover such activities by independent entities within the executive branch.)			James Bickley, 7-7794
Federal Claims Collection Act of 1966; Debt Collection Act of 1982; Debt Collection Improvement Act of 1996 (III.M., III.N., and III.P. in RL30795) (These statutes have been combined because the 1982 and 1996 acts amend the 1966 act.)			X (The statutes broadly cover "executive, legislative, and judicial agencies," but it is unclear whether this includes elements of the EOP.)	X (The statutes broadly cover "executive, legislative, and judicial agencies," so they appear to cover independent entities within the executive branch.)			Thomas Nicola, 7-5004

Table 1. (Continued).

General Management Law (location in CRS Report RL30795)	If the Office of the National Intelligence Director (NID) and National Counterterrorism Center (NCTC) were established ...						CRS Analyst, Telephone Number
	... in the Executive Office of the President (EOP), the NID and the NCTC appear to be:			... as independent entities, outside the EOP, but in the executive branch, the NID and the NCTC appear to be:			
	Covered	Not Covered	Not Clear	Covered	Not Covered	Not Clear	
Federal Debt Collection Procedures Act of 1990 (III.O. in RL30795)			X (This statute broadly covers any "agency, department, commission, board, or other entity of the U.S. or an instrumentality of the U.S.," but it is unclear whether this includes elements of the EOP.)	X (This statute broadly covers any "agency, department, commission, board, or other entity of the U.S. or an instrumentality of the U.S.")			Thomas Nicola, 7-5004
Improper Payments Information Act of 2002 (III.Q. in RL30795)			X (This act covers "executive agencies" per 31 U.S.C. § 102, which broadly defines the term as a "department, agency, or instrumentality" of the U.S. government, but it is not clear whether this includes the EOP.)	X (This act covers "executive agencies" per 31 U.S.C. § 102, which broadly defines the term as a "department, agency, or instrumentality" of the U.S. government.)			Virginia McMurtry, 7-8678
Cash Management Improvement Act of 1990 (III.R. in RL30795)			X (This act covers "executive agencies" per 31 U.S.C. § 102, which broadly defines the term as a "department, agency, or instrumentality" of the U.S. government, but it is not clear whether this includes the EOP.)	X (This act covers "executive agencies" per 31 U.S.C. § 102, which broadly defines the term as a "department, agency, or instrumentality" of the U.S. government.)			Steven Maguire, 7-7841

Table 1. (Continued).

General Management Law (location in CRS Report RL30795)	If the Office of the National Intelligence Director (NID) and National Counterterrorism Center (NCTC) were established ...						CRS Analyst, Telephone Number
	... in the Executive Office of the President (EOP), the NID and the NCTC appear to be:			... as independent entities, outside the EOP, but in the executive branch, the NID and the NCTC appear to be:			
	Covered	Not Covered	Not Clear	Covered	Not Covered	Not Clear	
User Fee Act of 1951 (III.S. in RL30795)			X (This act covers "executive agencies" per 31 U.S.C. § 102, which broadly defines the term as a "department, agency, or instrumentality" of the U.S. government, but it is not clear whether this includes the EOP.)	X (This act covers "executive agencies" per 31 U.S.C. § 102, which broadly defines the term as a "department, agency, or instrumentality" of the U.S. government.)			Frederick Kaiser, 7-8682
Government Corporation Control Act (IV.A. in RL30795)		X (This act (31 U.S.C. § 9101) covers government corporations, which does not include entities within the EOP.)			X (This act (31 U.S.C. § 9101) covers government corporations, which does not include independent entities in the executive branch.)		Kevin Kosar, 7-3968
Reorganization Act of 1977, as Amended (IV.B. in RL30795)		X (Reorganization authority expired in 1984 and would therefore not be in effect.)			X (Reorganization authority expired in 1984 and would therefore not be in effect.)		Henry Hogue, 7-0642
Federal Vacancies Reform Act of 1998 (IV.C. in RL30795)	X (The act (5 U.S.C. § 3345) covers all advice and consent positions, and specifically mentions the EOP.)			X (The act (5 U.S.C. § 3345) covers all advice and consent positions.)			Morton Rosenberg, 7-7480

Table 1. (Continued).

General Management Law (location in CRS Report RL30795)	If the Office of the National Intelligence Director (NID) and National Counterterrorism Center (NCTC) were established ...						CRS Analyst, Telephone Number
	... in the Executive Office of the President (EOP), the NID and the NCTC appear to be:			... as independent entities, outside the EOP, but in the executive branch, the NID and the NCTC appear to be:			
	Covered	Not Covered	Not Clear	Covered	Not Covered	Not Clear	
Public Buildings Act (V.A. in RL30795)			X (The act covers "executive agencies," which is defined (40 U.S.C. § 3301) as an executive department or independent establishment in the executive branch of the federal government. It is unclear whether this includes the EOP.)	X (The act covers "executive agencies," which is defined (40 U.S.C. § 3301) as an executive department or independent establishment in the executive branch of the federal government.)			Stephanie Smith, 7-8674
Federal Acquisition Streamlining Act (V.B. in RL30795)			X (The act (40 U.S.C. § 102) defines an executive agency as an executive department or independent establishment, or a wholly owned government corporation. It does not indicate that the EOP is included.)	X (The act (40 U.S.C. § 102) defines an executive agency as including independent establishments.)			Stephanie Smith, 7-8674
Federal Activities Inventory Reform (FAIR) Act of 1998 (V.C. in RL30795)			X (The act defines covered "agencies" by referring to 5 U.S.C. § 104 on independent establishments, but excludes the Government Accountability Office (GAO), government corporations, and others. It is unclear if an entity within the EOP is considered an "independent establishment.")	X (The act defines covered "agencies" by referring to, e.g., 5 U.S.C. § 104 (GAO and independent establishments other than the Postal Service and the Postal Rate Commission), but specifically excludes GAO, government corporations, and others. Therefore, if the proposed entities were independent agencies, they would appear to be covered by the act.)			Elaine Halchin, 7-0646

Table 1. (Continued).

General Management Law (location in CRS Report RL30795)	If the Office of the National Intelligence Director (NID) and National Counterterrorism Center (NCTC) were established ...						CRS Analyst, Telephone Number
	... in the Executive Office of the President (EOP), the NID and the NCTC appear to be:			... as independent entities, outside the EOP, but in the executive branch, the NID and the NCTC appear to be:			
	Covered	Not Covered	Not Clear	Covered	Not Covered	Not Clear	
Services Acquisition Reform Act (SARA) of 2003 (V.D. in RL30795)			X (The act defines the term "executive agency" by referring to 5 U.S.C. § 101, § 102, and § 104(1) (independent establishments), and wholly owned corporations subject to Chapter 91of Title 31. It is unclear if an entity within the EOP would be considered an "independent establishment.")	X (The act defines the term "executive agency" by referring to 5 U.S.C. 104(1) (independent establishments), which appears to include independent agencies in the executive branch.)			Elaine Halchin, 7-0646
Competition in Contracting Act (V.E. in RL30795)			X (The act (41 U.S.C. § 252) directs "executive agencies" to make purchases and contracts for services in accordance with its provisions. It is not clear whether this definition includes the EOP.)	X (The act (41 U.S.C. § 252) directs "executive agencies" to make purchases and contracts for services in accordance with its provisions.)			Michael Seitzinger, 7-7895
Davis-Bacon Act (V.F. in RL30795)	X (The act generally covers all contracts in excess of $2,000 to which the federal government is a party. However, the act may not cover instances when funds are channeled through a nonfederal entity (e.g., a state revolving fund). Also, the President can suspend the act during a "national emergency" (not defined).			X (The act generally covers all contracts in excess of $2,000 to which the federal government is a party. However, the act may not cover instances when funds are channeled through a non-federal entity (e.g., a state revolving fund). Also, the President can suspend the act during a "national emergency" (not defined).			William Whittaker, 7-7759; Morton Rosenberg, 7-7480

Table 1. (Continued).

General Management Law (location in CRS Report RL30795)	If the Office of the National Intelligence Director (NID) and National Counterterrorism Center (NCTC) were established ...						CRS Analyst, Telephone Number
	... in the Executive Office of the President (EOP), the NID and the NCTC appear to be:			... as independent entities, outside the EOP, but in the executive branch, the NID and the NCTC appear to be:			
	Covered	Not Covered	Not Clear	Covered	Not Covered	Not Clear	
Walsh-Healey Act (V.F. in RL30795)			X (The act (41 U.S.C. § 35) indicates that it covers (among others) any "agency or instrumentality of the United States." However, it may not be fully clear whether entities within the EOP are covered.)	X (The act covers any "agency or instrumentality of the United States," so it would likely cover the proposed entities. However, the act does not cover certain types of purchases, and may not cover instances when funds are channeled through a non-federal entity or when requirements "will seriously impair the conduct of Government business.")			William Whittaker, 7-7759; Morton Rosenberg, 7-7480
McNamara-O'Hara Service Contract Act (V.F. in RL30795)	X (The act (41 U.S.C. § 351) generally covers all contracts in excess of $2,500 entered into by the United States, "the principal purpose of which is to furnish services in the United States through the use of service employees."			X (The act (41 U.S.C. § 351) generally covers all contracts in excess of $2,500 entered into by the United States, "the principal purpose of which is to furnish services in the United States through the use of service employees."			William Whittaker, 7-7759; Morton Rosenberg, 7-7480
Prompt Payment Act (V.G. in RL30795)	X (The definition of "agency" in the act refers to the definition in the APA (5 U.S.C. § 551(1)), which covers each authority of the government, "whether or not it is within or subject to review by another agency." As with the APA, however, if either organization were strictly advisory in nature, it would not be covered.)			X (The definition of "agency" in the act refers to the definition in the APA (5 U.S.C. § 551(1)), which covers each authority of the government, "whether or not it is within or subject to review by another agency." As with the APA, however, if either organization were strictly advisory in nature, it would not be covered.)			John Luckey, 7-7897

Table 1. (Continued).

General Management Law (location in CRS Report RL30795)	If the Office of the National Intelligence Director (NID) and National Counterterrorism Center (NCTC) were established ...						CRS Analyst, Telephone Number
	... in the Executive Office of the President (EOP), the NID and the NCTC appear to be:			... as independent entities, outside the EOP, but in the executive branch, the NID and the NCTC appear to be:			
	Covered	Not Covered	Not Clear	Covered	Not Covered	Not Clear	
Ethics in Government Act (VII.B. in RL30795)	X (The financial disclosure requirements in the act (5 U.S.C. App.) generally cover each officer or employee in the executive branch above GS-15 in the General Schedule (or equivalent).)			X (The financial disclosure requirements in the act (5 U.S.C. App.) generally cover each officer or employee in the executive branch above GS- 15 in the General Schedule (or equivalent).)			Mildred Amer, 7-8304
Ethics Reform Act of 1989 (VII.C. in RL30795)	X (The act amended the Ethics in Government Act, so its coverage is the same as above.)			X (The act amended the Ethics in Government Act, so its coverage is the same as above.)			Mildred Amer, 7-8304
Lobbying with Appropriated Monies Act (VII.D. in RL30795)	X (The act (18 U.S.C. § 1913) generally covers "money appropriated by any enactment of Congress." Therefore, it appears to cover entities within the EOP as long as they receive appropriated funds.)			X (The act (18 U.S.C. § 1913) generally covers "money appropriated by any enactment of Congress." Therefore, it appears to cover independent entities within the executive branch as long as they receive appropriated funds.)			Louis Fisher, 7-8676
Federal Tort Claims Act (VII.E. in RL30795)			X (The act (28 U.S.C. § 2671) covers a wide range of entities, including "executive departments" and "independent establishments of the United States," but is not clear about whether the EOP falls within these categories.)	X (The act (28 U.S.C. § 2671) covers "independent establishments" of the U.S. government.)			Henry Cohen, 7-7892

Evaluating Options: Standardization vs. Customization

When considering whether these general management laws should cover the proposed NID and NCTC, if any, or whether alternative or new management-related laws should be enacted, Congress might consider an issue that it has previously confronted when reorganizing or creating agencies — whether all (or virtually all) these management laws should cover all (or virtually all) agencies, or whether each agency should be governed by a customized set of these laws. Some assert that the laws should cover as many agencies as possible, arguing that less coverage would lessen the laws' importance and influence, and would potentially make it more difficult for Congress to hold the President and agencies accountable for their administration. Others contend that each agency is sufficiently different to merit individualized attention, and that there should be no presumption of coverage with regard to any specific agency — particularly with regard to intelligence agencies or other entities that deal with sensitive or classified information. For analysis and discussion of potential advantages and disadvantages of approaches to these decisions, see CRS Report RL32388, *General Management Laws: Major Themes and Management Policy Options*, pp. 13-15.

REFERENCES

[1] U.S. National Commission on Terrorist Attacks Upon the United States, *The 9/11 Commission Report* (Washington: GPO, July 22, 2004), p. 399.

[2] Ibid., p. 411.

[3] Ibid., p. 412.

[4] Ibid., p. 414.

[5] Ibid., p. 403.

[6] Ibid., p. 405.

[7] This chapter identifies legislation introduced or announced as of Sept. 15, 2004. For a discussion of this debate, see CRS Report RL32506, *The Position of Director of National Intelligence: Issues for Congress*, by Alfred Cumming. See also CRS Report RL32501, *9/11 Commission Recommendations: New Structures and Organization*, by Harold C. Relyea.

[8] See, for example, S. 2645 (107th Cong.), introduced June 19, 2002. For legislation in the 108th Congress, see S. 6, introduced on January 7, 2003; S. 190, introduced on January 17, 2004; S. 1520, introduced on July 31, 2003; H.R. 4104, introduced on April 1, 2004; and H.R. 4584, introduced on June 16, 2004. For a discussion of these legislative proposals, see CRS Report RL32506, *The Position of Director of National Intelligence: Issues for Congress*, by Alfred Cumming.

[9] According to a news report, the 9/11 Commission's vice chairman, Lee H. Hamilton, "said he withdrew the panel's proposal that the new national intelligence director be part of the executive office of the president, citing opposition from both the White House and Congress." Helen Dewar and Charles Babington, "Intelligence Retooling on Agenda as Congress Returns," *Washington Post*, Sept. 8, 2004, p. A4.

[10] Ibid.

[11] For the press release, see [http://www.senate.gov/~gov_affairs/index.cfm?Fuseaction=PressReleases.DetailandPressRelease_id=796andMonth=9andYear=2004andAffiliation=R], visited Sept. 20, 2004.

[12] See [http://www.senate.gov/~govt-aff/_files/091504billsummary.pdf], p. 2, visited Sept. 20, 2004.

[13] Executive Order 13355, "Strengthened Management of the Intelligence Community," 69 *Federal Register* 53593, Sept. 1, 2004.

[14] Executive Order 13354, "National Counterterrorism Center," 69 *Federal Register* 53589, Sept. 1, 2004.

[15] CRS Report RL30795, *General Management Laws: A Compendium*, coordinated by Clinton T. Brass; and CRS Report RL32388, *General Management Laws: Major Themes and Management Policy Options*, by Clinton T. Brass.

[16] Some laws have provisions in which determinations of coverage are mixed. For example, in the EOP scenario, the Clinger-Cohen Act of 1996 has some provisions that would not cover the proposed entities, but also has provisions in which coverage is not clear. Conversely, in the independent agencies scenario, some provisions of the act would appear to cover the proposed entities, but others would not.

[17] 5 U.S.C. § 551(1).

[18] *Soucie v. David*, 448 F.2d 1067, 1073 (1971).

[19] 5 U.S.C. § 552(f)(1).

[20] *Meyer v. Bush*, 981 F.2d 1288, 1297 (D.C. Cir. 1993).

[21] This standard is likewise applicable to the Privacy Act, which incorporates the definition of "agency" as used in FOIA.

[22] GPRA is codified at 5 U.S.C. prec. § 301, § 306; 31 U.S.C. § 1101 and nt, §1105, §§ 1115-1119, prec. § 9701, §§ 9703-9704; 39 U.S.C. prec. § 2001, §§ 2801-2805. If precedent were followed, it is possible that the NID and the NCTC would be excluded from GPRA's coverage. According to the 9/11 Commission report, the NCTC would "be built on the foundation of the existing Terrorist Threat Integration Center" (p. 403), which is located in the CIA. The CIA was excluded from GPRA's coverage by 5 U.S.C. § 306(f). According to GPRA's legislative history, the CIA was excluded largely because most of its budget is classified, and because the GPRA plans and reports it would prepare would be classified (S. Rep. 103-58, *Government Performance and Results Act of 1993*). OMB reported to Congress in 1997 that "OMB consulted with the Agency leadership about this matter, and the CIA agreed to comply voluntarily with the basic provisions of GPRA" (*The Government Performance and Results Act, Report to the President and Congress from the Director of the Office of Management and Budget*, May 1997.) Congress subsequently enacted P.L. 104-293, the Intelligence Authorization Act for FY1997, to require performance reports on intelligence activities, requirements, and evaluations (Secs. 803 and 808 of P.L. 104-293).

[23] As alternatives, see CRS Report RL30795, *General Management Laws: A Compendium*, coordinated by Clinton T. Brass, or call the CRS experts listed in the table.

In: Computer Networks, Policies and Applications
Editors: C. M. Berger and A. D. Streusse
ISBN: 978-1-61209-088-7

Chapter 8

THE ICT'S ACCEPTANCE: ANALYTICAL FRAMEWORK AND EMPIRICAL EXAMPLE

Jamel Choukir[1], Mouna Baccour Hentati[1], Bilel Bellaj[1] and Hichem Eleuch[2,*]
[1]University of Sfax,
Unité de Recherche Gouvernance, ESC BP 1081, Sfax, Tunisia
[2]Institut National des Sciences Appliquées et de Technologie
BP 676, Zone Urbaine Nord, 1080 Tunis, Tunisia

ABSTRACT

ICT are challenge for the Tunisian economy. This study highlights the acceptance as a social issue that can helps to understand the behaviour of student toward ICT. A sample of 200 Tunisian students was used for the survey. The results show the links between motivator factors and acceptance and on the other hand some ties between acceptance, motivator factors and the antecedents, especially gender, age and study level. According to our results some factors seem to be more important for students such as the perceived usefulness and the perceived enjoyment. Social pressure and perceived ease of use are considered as weak factor to understand the acceptance, exception for student between 20 and 22 years; the results show some differences.

Keywords: acceptance; beliefs, creation and personal development, perception, ICT, Social impact.

1. INTRODUCTION

Data show that ICTs[1] purchases are increasing everywhere; microcomputers purchases in Tunisia, for example, are growing from 400.3 thousand (2003) to 472.1 thousand (2004);

---

* E-mail address: hichemeleuch@yahoo.fr

Internet users are passing from 600 thousand (2003) to 953.8 thousand (2005). Concerning the ICTs use, Tunisia was ranked 31th among 120 countries [1], these facts don't necessarily mean that ICTs are well utilized; hence they make significant contribution to improve individual and collective performance.

With an increasing interest in the use of ICTs in education and training, the concern about just how our students react and compose with. A quick show of hands for how many people know any particular application, let us say Word, Excel, Access, Power Point, we reach a result close to unanimous response of yes. But using the program and knowing it are very different things.

Understanding the factors that determine acceptance of information and communication technologies is of interest both to researchers in many fields and to practitioners. This paper reviews literature which demonstrates that the technology acceptance was related to the technology characteristics such as: relative advantage, compatibility; complexity and observability. Some authors [2] and [3] consider that the diffusion of the ICTs was the consequence of these characteristics. Others extend the diffusion to more complex factors such as the managerial influence (reward system and incentives). Our research try to understand the antecedent of student's technology acceptance; in other words how student users compose with ICTs. Davis [2] through his Technology acceptance model (TAM), explains that the user attitudes were determined by two factors: perceived ease of use and perceived usefulness. The TAM model postulates that ease of use and usefulness are significant determinants to ICTs use. Shackel's [4] acceptability system presents three dimensions: utility + usability +likeability. We consider that the acceptance has been viewed as a function of demographic variables (age, gender and education), social norms (social roots and localisation) and user perceptions (beliefs and trust).

By understanding the antecedents of ICTs acceptance, we elucidate information about successful ICTs conditions, the process for better use and developing added value. The main purpose of this study is to assess the pertinence of two models usually used. Our research is a confirmatory one; we attempt the validation of TAM and TRA models in the Tunisian context.

2. Background of the Study

Davis [2] considers user acceptance as the pivotal factor determining the success or failure of any information system project. In fact, lack of user acceptance is a significant barrier to the success of new information systems. We argue that ICTs acceptance by users was the Success Key Factor. But it can be deviated from the tasks that it designed to support. We advance that the ICTs success depends on its acceptability, which means not a simple use but active and proactive involvement.

For the present purposes, acceptance is defined as attitude that implies adaptation and integration of information technology, as example in the education and training field. Positive attitudes mean that ICTs can allow students to create added value. We distinguish a simple use or adoption from acceptance which means involvement, willingness and beliefs in the pertinence of ICTs and their appropriation.

[1] ICTs mean all kinds of computers, applications and networks

3. Conceptual Model

Our Review of the empirical studies in this area shows that the theory of reasoned action (TRA) and the theory of acceptance model (TAM) are the most popular theories used to explain the use, the adoption and acceptance. These models constitute a starting point to understand the engine of acceptance as attitude which, we advance that, lead to personal development.

Shackel's [4] acceptability paradigm provides a useful theoretical starting point for this paper. It is the most influential paradigm for conceptualizing the acceptability of any given system to its intended users. Shackel suggests that acceptability can be defined as a function of three orthogonal dimensions utility (whether the system does what is needed functionally), plus usability (whether the users can actually work the system successfully), plus likeability, (whether the users feel the system is suitable, all balanced against cost, which includes both capital and operating expenses as well as social consequences of system use).

In the literature review, the TAM model is considered as the most influential contemporary IT research paradigm. It is important to know that the conceptual framework established by TAM enjoyed validation across users, technologies and task settings in academic and professional communities [2], [5], [6] and [7]. The TAM model considers that user behaviour depends on the perceptions of usefulness and the perceptions of ease of use. With the TAM model, perceived usefulness and perceived ease of use are considered as the determinant of attitudes [2]. Davis' investigation showed that perceived usefulness and ease of use were significantly related to self-report of usage behaviour, although perceived usefulness was strongly related to usage. Results of Davis' study indicated that the relative importance of ease of use and usefulness varies as a function of time.

Other authors suggest that the mechanisms determining user behavior are more complex than proposed by the TAM model. For example, Adams et al [6] suggested that their findings might be better explained if perceptions of ease of use varied as a function of user experience level. In an attempt to measure the magnitude of these temporal effects, recent research efforts have proposed a "revised" TAM model which addresses how the interrelations between usefulness and ease of use, attitudes intentions and users behaviours vary as a function of time or experience [8-9]. Specifically, Chau [10] indicated that ease of use had a strong effect on perceptions of short-term usefulness, but no direct effect on intention or long-term usefulness.

According to the theory of reasoned action, we can consider that behaviour, in this instance ICTs acceptance, is influenced by individual perceptions and attitudes as well as by social influences. Deci's [11] distinguishes between intrinsic and extrinsic motivators of behaviours. he considers that individuals may be motivated to use ICTs because of the intrinsic rewards derived (enjoyment and fun) perceived benefits (usefulness) or external pressure (social pressure). To elucidate the ICTs acceptance, we adopt the TRA and TAM a revised models as a framework.

3.1. The Impact of ICTs Acceptance

Information technology has been mainly studied in terms of advantages. Most researchers emphasis on their benefits and efficiency. On the one hand, ICTs increase quality, productivity and reduce costs; on other hand, Biskson et al. [12] and Porter [13] reported that with, as example, office automation, middle managers found their job more enriching and the quality of work life enhanced. However, contradictory findings have been reported about ICTs. Some studies [14] showed that ICTs may increase productivity but it can also degrade the work and social life of users. Kraut et al. [15] added that the individual users quality of work life became poor and had less variety. In fact, information technology was not neutral; advantages or disadvantages parts depend on users profile and the context of use.

3.2. Motivational Factors

According to our literature review, we consider that students acceptance of ICTs depends on four motivators: The perceived ease of use, the perceived usefulness, the perceived enjoyment and the social pressure. Perceived benefits do not mean necessarily materiel but it can be psychological one. The perception is cognitive vocabulary which is mental representation implying some behaviour.

3.2.1. The Perceived Ease of Use

The perceived ease of use is defined, by Davis [2] as "the degree to which a person believes that using a particular system would be free of effort". According to Davis, ease of use is considered as an important determinant of system usage through perceived usefulness. We think that ICTs are simple for use, which theoretically increases the number of user.

3.2.2. The Perceived Usefulness

The perceived usefulness is defined as "the degree to which a person believes that using a particular system would enhance his or her job". It's considered as an important motivating factor. It derives from the TRA and TAM models. According to these models, the perceived usefulness affects ICTs acceptance or use. The more beneficial (commodity and simplicity) you think that the ICTs vehicle the more acceptance you get.

3.2.3. The Perceived Enjoyment

The acceptance of ICTs can also be motivated by intrinsic psychological incentives and rewards. Fun and pleasure represent an intrinsic motivation for ICTs acceptance. Empirical studies show that the general characteristics of enjoyment/fun relate positively to creativity and exploratory type of behaviour during interaction with computers [16]. Davis et al. [2] found that, while perceived usefulness emerged as the major determinant of computer acceptance in the workplace, enjoyment and fun had a significant effect beyond perceived usefulness. The fun as a feeling creates positive behaviour toward ICTs.

3.2.4. The Social Pressure

According to the TRA model, in addition to the individual perceptions and beliefs, social influences can affect behaviour [17]. Social pressure can affect the attitudes and behaviour of individuals in different degrees depending on societies and cultures. Social pressure represents not only the incentives (reward) but also the constraints (obligation by registration, courses, homework and exams).

Proposition: The ICTs acceptance is a function of four motivational factors: the perceived ease of use, the perceived usefulness, the perceived enjoyment and the social pressure. However, these factors do not act equally and in the same degree. Their effects depend on the social and demographic profile (the gender, the age and the level and discipline).

3.3. Antecedents Factors

We consider that there are two kinds of levels: the individual and the organizational one, considered as the antecedents for ICTs students acceptance.

3.3.1. The Individual Level

Three variables were identified as individual antecedents: gender, age and education.

The Gender

Many data [1] show that access to the ICTs is related to the gender; it seems that men enjoy of more access opportunities than women.

Female students represent about 65% of all students in Tunisian. This data indicates that this presence was a sign of Tunisian education system success, but nothing can be concluded about the attitudes toward ICTs. The lack of information about the link pushes us to advance that gender can be a significant determinant of ICTs acceptance; it can also be considered as a factor affecting the modalities of use; access can be considered as a first step for acceptance.

The Age

Data show that the majority of Internet users are students. In fact, age, as a demographic variable, can be a significant determinant of the ICTs acceptance. For students, we think that we do not need to demonstrate that all this population was convinced that the ICTs were not luxurious but a must. We think the younger you are the more ICTs users we find and the more positive attitudes we will obtain.

The Education

Education is one of the important variables determining ICTs acceptance. We think that the frequency (experience) and the intensity of use depend on the nature of student discipline; certainly, ICTs can be accepted and adopted by all students and may be considered as success factor key. Both the discipline and the study level can be considered as factors determining the ICTs acceptance.

3.3.2. The Organizational Level

The organizational level refers to the support and the incentives can facilitate the emergence of positive attitudes towards ICTs. The positive link between organizational support and system usage is documented in the literature [18-19]. The lack of organizational support can be considered as a limitation to the ICTs acceptance.

4. Methodology

The investigation was conducted both by a documentary analysis, concerning the organizational support and by a survey.

4.1. The Sample

Our sample was composed of students from the university of Sfax, the second most important university in the country (50000 students; 20% of the total number of Tunisian students). The students belong to all discipline; we chose 20 students from every discipline.

The data for this study were gathered through means of a survey questionnaire administered to 200 students in French language. The survey contains an assignment explaining the purpose of the study. The survey was administered, the third week of April, 2007. We realized this survey ourselves directly with no intermediary, but in some cases, we mobilized our social capital; it means our colleagues, friends and students. Of the 200 surveys, 180 were completed and returned. However, only 160 usable surveys were retained for data analysis, providing a response rate of 80%. The period of administration of our survey corresponding to last week before the exam preparation which explain this rate.

4.2. The Survey Structure

We define four dimensions: the antecedents, motivator factors, acceptance and personal development. Each dimension was explored by some items. Most of the responses (17/25) were anchored in seven-point scales. Individuals were asked to indicate the extent of agreement or disagreement on a seven-point Likert-type scale ranging from (strongly disagree) to (strongly agree).

4.2.1. The Antecedents

We distinguish two types of levels the individual and the organizational one. The individual level was explored through these variables: age, gender and education (level and discipline). The organizational one concerns the support provided by management (resources) to get individual to use the ICTs quickly.

4.2.2. The Motivator Factors

According to our model research, we identified four motivational factors.

Perceived ease of use was explored through theses items:

- Learning to use ICTs would be easy for me.
- I would find it easy to get ICTs to do what I want to do.
- It would be easy for me to become skilful using ICTs and I would find ICTs easy to use.

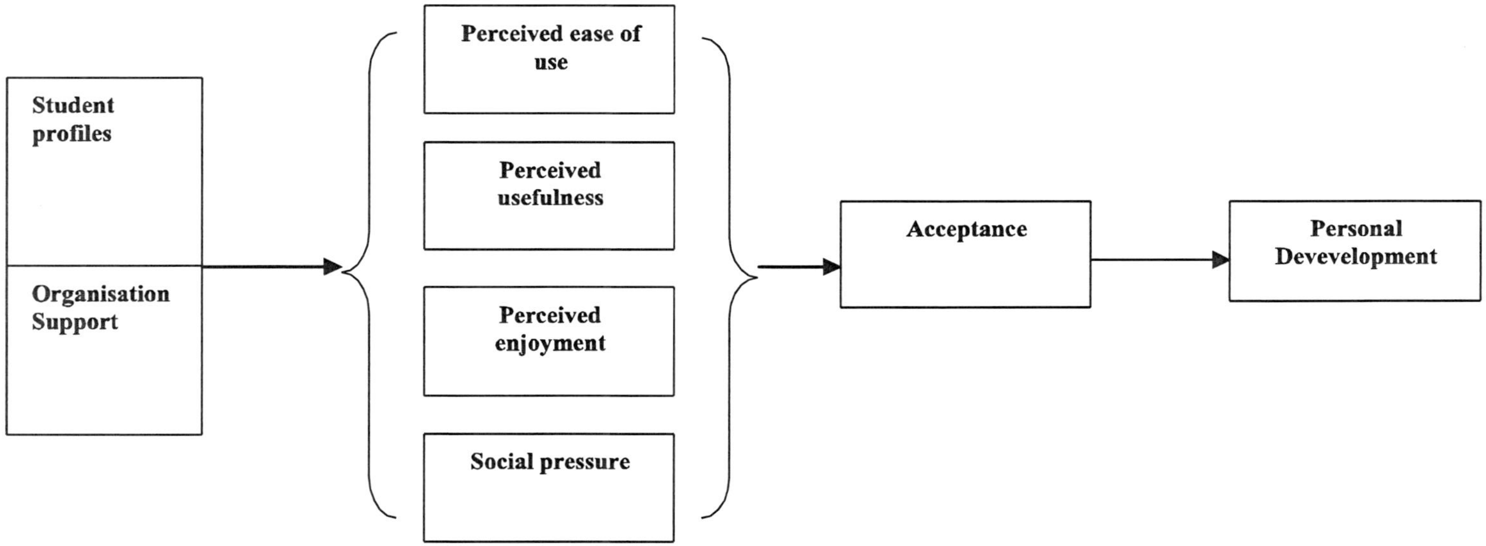

Figure 1. Research Model.

Perceived usefulness was explored through theses items:

- Using ICTs improves my learning ability.
- Using ICTs could provide me with information that would lead to better learning.
- Using ICTs would lead to personal development.

Perceived enjoyment was explored through theses items:

- Using ICTs in my learn is rewarding.
- Using ICTs in my learn is pleasurable.
- Using ICTs in my learn is beneficial.

Social pressure was explored through theses items:

- Teachers in general think I should use ICTs in my learning.
- Students think we should use ICTs regularly.
- Managers think that our chance to get a job depends on our ability to manipulate ICTs.
-

4.2.3. Acceptance

Acceptance was explored through theses items:

- The frequency of use is measured through the experience and the time.
- The types or the kind of applications used.
- Beyond these items, acceptance can be measured through intensity and the kind of use.

4.2.4. Personal development

Personal development was explored through these items:

- Using ICTs increases my ability to learn.
- Using ICTs enables me to learn.
- Using ICTs keeps me some distinction.

4.3. Data Analysis

In The first step, we generate the descriptive analysis in term of frequencies and profile.

In the second step, we precede by some tests such as *the fiability analysis* (α of Crombach) and *unidimensionality analysis* (factor reduction analysis with varimax rotation), to measure the scale pertinence. We have to obtain a fiability superior to 0.6 [20]. This test allowed as the respecting of the most important condition for the data treatment. We obtained for the first construct (Perceived ease of use), one factor (62,258%) respecting the one-dimensional of the selected measures and α of Crombach = 0.8006.

For the second construct (Perceived usefulness), we obtained one factor (80.49%) and α of Crombach = 0.7534. For the third construct (Perceived enjoyment), we obtained two factors (70.612%), in fact the unidimensionality is not respected and α of Crombach = 0.517 (unacceptable).

Finally, for social pressure; the factor reduction analysis generates one factor (63.38%) and for the second construct (Perceived usefulness), we obtained one factor (80.49%) and α of Crombach = 0.7362. In the end, we proceed by linear regression which means to explain ITCs acceptance through motivational factors by the integration of some antecedents (socio-demographic variables).

Table 1. Socio-demographic profile

Characteristics	**Modalities**							
Gender	*Male 43%*				*Female 56%*			
Age	*18 à 20 years 15.6%*				*20 à 22 years 41.3%*		*More22 years 43.1%*	
Discipline	*Management 22.5%*	*Engineer 12.5%*	*Medical 10.6%*	*Health 6..9 %*	*Arts 12.5%*	Music 12.5 %	Literature 10.6%	Law 11.9%
Level	undergraduates 45.6%		graduates 35.6%			postgraduates 18.8%		

5. Discussion

Our results can be summarized as following:

5.1. Factors Affecting Attitudes toward ICTs Acceptance

Results on perceived consequence are interesting; the explanation power shows the following order, from the most important to less important factor: the perceived usefulness, the perceived enjoyment, the perceived ease of use and finally the social pressure.

Results on Perceived Ease of Use

This factor explain the acceptance in term of R squared equal to 24% (F Fisher = 44 which is an interested explanation power) and the coefficient β equal to 0.5 (significant student T equal to 6.7).

Results on Perceived Usefulness

This factor explain the acceptance in term of R squared equal to 29% (F Fisher = 57 which is an interested explanation power) and the coefficient β equal to 0.54 (significant student T equal to 7.6).

Results on Perceived Enjoyment

This factor explain the acceptance in term of R squared equal to 27% (F Fisher = 51 which is an interested explanation power) and the coefficient β equal to 0.52 (significant student T equal to 7.2).

Results on Social Pressure

This factor explain the acceptance in term of R squared equal to 17% (F Fisher = 28, which is a less interested explanation power and the coefficient β equal to 0.4 ((significant student T equal to 5.3).

According to these results, we note that the motivator factors did no act equally; some are most important especially the perceived usefulness and perceived enjoyment. We argue that the explanation lies in the characteristic of the studied population. The pleasure and the economical aspects constitute a determinant motivator factor. We think that the perceived usefulness and the perceived enjoyment were related. Each factor reinforces the other.

5.2. Factors Affecting ICTs Acceptance

The PCA results allowed us to convert the items on factors. To apply the linear regression (ANOVA) we obtained the followed model: We obtain this equation after testing the non-multicolinearity of the motivator factors.

$$Y= C +a_1 F1 +a_2 F2 + a_3 F3+ a_4 F4$$

where a_i represent the coefficients of linear regression equation.

The third construct $F3_1$ and $F3_2$

$$Z= C+a_1 F1 +a_2 F2 + a_{3\text{-}1} F31 +a_{3\text{-}2} F32 +a_4 F4$$

We obtained for the global model significant results as following an R squared =0.4 with an F of Fisher[1] =11.83 › 2 and the T of Student for each coefficient a_i (that are the output of SPSS 11.0). The details of the equation are:

This means that the motivator factors contribute with about 40% for the explication of the ICTs acceptance measured by the intensity use. The most significant factors were the perceived usefulness and the perceived enjoyment with a significant student T. In second step, we introduced the antecedent factors as filter variables. This analysis has led to the following results:

Table 2. Linear Regression for global model

Model regression results

Model	R	R-squared	AdjustedR-squared	Standard error of estimation
1	,605	,366	,343	,855

[1] Ficher: explicative power of the model T of Student : Degree of significativity of coefficients (**a**i).

Table 3. Estimation of the model
Predict values: C, a_4, a_{32}, a_{31}, a_1, a_2 Coefficients

		Non- standard Coef.		Coef. non standard.	T	Signif.
Model		B	Standard Error	a_i		
1	C	1,737	,105		16,486	,000
	a_1	,128	,096	,134	1,337	,183
	a_2	,226	,096	,251	2,353	,020
	a_{31}	,223	,082	,249	2,706	,008
	a_{32}	-7,039E-02	,063	-,078	-1,120	,265
	a_4	5,764E-02	,084	,061	,684	,495

Results on Gender

For male students the explicative power increases (0,313 against 0,425). For female students the determinant motivator factor was the perceived enjoyment. For men this factor act negatively. However, for men the most important factor was the perceived usefulness.

Results on Age

Two age groups share the same motivation degree (R squared equal to 0.5). It means that the significant motivator factor concern the perceived ease of use for students between 18 and 20. For students those ages were more than 22, the perceived enjoyment and the perceived usefulness constitute a significant factor of motivation. The result was different for students between 20 and 22. The perceived usefulness affects negatively the ICTs acceptance. However for the connected people, the perceived enjoyment (0,434) was a determinant motivator factor and also the social pressure (0,332) (significant student T for the two factors).

Results on Education (Level and Discipline)

For postgraduate's students the R squared reaches the maximum of 0,84. This result constitutes a validation of our model research. The perceived enjoyment is the significant factor. For graduate's students the R squared decreased and reaches the 0,35. The determinant factors were in first place perceived ease of use and perceived usefulness in second place. The R squared for undergraduate's students was 0.25. The more affected motivator factor was the perceived enjoyment and in second place the perceived usefulness.

CONCLUSION

The present study integrated the theoretical perspectives and empirical findings of research on the ICTs acceptance and its impacts on development performance. A model is proposed and tested examining the role of the motivator factors on ICTs acceptance and the role of antecedent factors on the motivator factors.

The results of this study indicate that the perceived usefulness, the perceived enjoyment constitute the most determinant ICTs acceptance motivator factors. The social pressure is shown as non significant motivator factor. The only exception was for students between 20 and 22 years. May be due to the fact that this category of people are learning to employment which can be manifested in more social pressure at this stage. In addition, the influence of these factors differs from men to women. For men the most significant factor was the usefulness contrary for women the enjoyment was the most determinant motivator factor. This difference can be explained by sociological features, i.e men are more pragmatic than women. This is not totally confirmed. It needs more validation. Still, more demonstration is needed in this respect.

The results of the study can contribute to a better theoretical understanding of the factors that promote the ICTs acceptance. They have far reaching implications for developed countries as well. The development and the acceptance require a salient understanding of the local culture. The societal dynamic and also the profile of ended users.

The results show the links between motivator factors and acceptance and on the other hand some ties between acceptance, motivator factors and the antecedents, especially gender, age and study level. According to our findings some factors seem to be more determinant for students such as the perceived usefulness and the perceived enjoyment. Social pressure and perceived ease of use can be considered as weak factor to understand the acceptance, exception for student between 20 and 22 years; the results show some differences. We present as a hypothesis that the enjoyment in one side and usefulness and social pressures on the other side are not necessarily antinomy, we oppose intrinsic motivation to extrinsic one, we advance that the first one can be considered as the most determinant for the acceptance.

A longitudinal research can be essential to confirm these links among the study variables. The strengths of the findings can be enhanced by extensive surveys (covering other regions and universities). Te first limitation concerns the sample which deals with only the University of Sfax. This research needs to be extended for other regions and cover all disciplines. However, the results show and give us interesting configuration of the ICTs acceptance; it allows asking many myths associated to the ICTs.

REFERENCES

[1] World Economic Forum Report (2006).

[2] Davis, F. D.. Perceived usefulness perceived ease of use and user acceptance of information technology. *MIS Quarterly*, **13**, 983-1003, (1989); Davis, F. D. Extrinsic and intrinsic motivation to use computers in workplace. *Journal of Applied Social Psychology,* **22**, 1111-1132, (1992).

[3] Rogers, E. M., *Diffusion of innovation*, New-York: The Free Press, (1995).

[4] Shackel, B. Usability-context, framework definitions, design and evaluation. In Shackel, B. and Richardson, S. Eds. *Human Factors for Informatics Usability*.pp.21-37. Cambridge, UK: Cambridge University Press (1991).

[5] Mathieson, K. Predicting user intentions: comparing the technology acceptance model with the theory of planned behaviour. *Information System Research*, **3**(3), 173-191, (1991).

[6] Adams, D. A. et al; Perceived usefulness ease of use and usage of information technology: A replication. *MIS Quarterly*, 16, 227-247, (1992).

[7] De Marez Lieven S.B., Verleye Gino B.M. ICT-innovations today: making traditional diffusion patterns obsolete, and preliminary insight of increased importance, *Telematics and Informatics* **21** (2004) 235–260.

[8] Taylor, S., and Todd, P. A. (1995). Assessing IT usage: the role of prior experience. *MIS Quarterly,* 19(4), 561-570; Taylor, S., and Todd, P. A. (1995). Understanding information technology usage: A test of competing models. *Information System Research*, **6**(2), 144-176 (1995).

[9] Szjana, B., Empirical evaluation of the revised technology acceptance model. *Management Science*, **42**(1), 85-92, (1996).

[10] Chau, P. K. Y., An empirical assessment of a modified technology acceptance model. *Journal of Management Information System*, **13**(2), 185-204, (1996).

[11] Deci's, E. L. *Intrinsic motivation*. New-York: Plenum Press,(1975).

[12] Biskson, T. K. et al., *Computer-mediated work: Individual and organizational impact in one corporate headquarters*. Santa Monica, Rand, (1985).

[13] Porter, A; L., A two factor model of the effects of office automation on employment. *Office: Technology and People*, **3**, 57-76, (1987)..

[14] Olson, J. M. and Zanna, M. P. Attitudes and attitude change. *Annual Review Psychology,* **44**, 117-154, (1993).

[15] Kraut, R. et al., Computerization, productivity, and quality of work-life. *Communications of the ACM*, **32**(2), (1989).

[16] Webster, J., Microcomputer playfulness: Development of a measure of workplace implications. *MIS Quarterly*, **16**(2), 201-226, (1992).

[17] Ajzen, I., *From intentions to actions: A theory of planned behaviour*. In J. Kuhl and J. Beckmann, *Action control: From cognition to behaviour*; New-York: Springer (1985).

[18] Fishbein, M., and Ajzen, I., *Belief attitude intentions and behaviour: An introduction to theory and research.* Boston: Addison-Wesley, (1975).

[19] Foster, D. and Cornford, T., Evaluation of health information system: Issues, models and case studies. *Social implications of computers in developing countries*. New-York: McGraw-Hill, (1992).

[20] Nunnaly, *Psychometric theory*. New-york, K. M. C. Graw Hill Book, (1967). forum report , 2003, 2004, 2005, 2006, 2007.

In: Computer Networks, Policies and Applications
Editors: C. M. Berger and A. D. Streusse
ISBN: 978-1-61209-088-7

Chapter 9

PRIVACY: GOVERNMENT USE OF DATA FROM INFORMATION RESELLERS COULD INCLUDE BETTER PROTECTIONS*

Linda D. Koontz

WHY GAO DID THIS STUDY

Federal agencies collect and use personal information for various purposes from information resellers—companies that amass and sell data from many sources. GAO was asked to testify on its April 2006 report on agency use of reseller data. For that report, GAO was asked to determine how the Departments of Justice, Homeland Security, and State and the Social Security Administration used personal data from resellers and to review the extent to which agencies' policies and practices for handling this information reflected the Fair Information Practices, a set of widely accepted principles for protecting the privacy and security of personal data. GAO was also asked to provide an update on the implementation status of its recommendations and to comment on provisions of the proposed Federal Agency Data Protection Act. In preparing this testimony, GAO relied primarily on its April 2006 report.

WHAT GAO RECOMMENDS

GAO is not making additional recommendations at this time. However, in its 2006 report, GAO made recommendations to the Office of Management and Budget and the four agencies to address agency use of personal information from commercial sources. Agency officials generally agreed with the content of the report. Since then, 2 of the 4 agencies have taken steps to address its recommendations; however, OMB has not issued clarified guidance.

* This is an edited, reformatted and augmented edition of a United States Government Accountability Office publication, Report GAO-08-543T, dated March 11, 2008.

What GAO Found

In fiscal year 2005, the Departments of Justice, Homeland Security, and State and the Social Security Administration reported that they used personal information obtained from resellers for a variety of purposes, including performing criminal investigations, locating witnesses and fugitives, researching assets held by individuals of interest, and detecting prescription drug fraud. The agencies planned spending approximately $30 million on contractual arrangements with resellers that enabled the acquisition and use of such information. About 91 percent of the planned fiscal year 2005 spending was for law enforcement (69 percent) or counterterrorism (22 percent).

Agency practices for handling personal information acquired from information resellers did not always fully reflect the Fair Information Practices. That is, for some of these principles, agency practices were uneven. For example, although agencies issued public notices when they systematically collected personal information, these notices did not always notify the public that information resellers were among the sources to be used. This practice is not consistent with the principle that individuals should be informed about privacy policies and the collection of information. Contributing to the uneven application of the Fair Information Practices are ambiguities in guidance from the Office of Management and Budget (OMB) regarding the applicability of privacy requirements to federal agency uses of reseller information. In addition, agencies generally lacked policies that specifically address these uses.

GAO made recommendations to OMB to revise privacy guidance and to the four agencies to develop specific policies for the use of personal information from resellers. The five agencies generally agreed with the report and described actions initiated to address the recommendations. Since GAO issued its report, agencies have taken steps to address the recommendations. For example, the Department of Homeland Security Privacy Office incorporated specific questions in its May 2007 Privacy Impact Assessment guidance concerning use of commercial data. In addition, the Department of Justice took steps to update its public notices to specify their use of data from information resellers. OMB, however, has not implemented GAO's recommendation to clarify guidance on use of commercial data.

The Federal Agency Data Protection Act was introduced on December 18, 2007. The legislation, among other things would require that agencies (1) conduct privacy impact assessments for their uses of commercial data, and (2) promulgate regulations concerning the use of commercial data brokers. GAO considers these requirements to be consistent with the results and the recommendations made to the agencies in its 2006 report.

Abbreviations

DEA	Drug Enforcement Administration
DHS	Department of Homeland Security
DOJ	Department of Justice
FBI	Federal Bureau of Investigation
OECD	Organization for Economic Cooperation and Development

OIG	Office of the Inspector General
OMB	Office of Management and Budget
PIA	privacy impact assessments
SSA	Social Security Administration
State	Department of State
TSA	Transportation Security Administration

Mr. Chairman and Members of the Subcommittee

I appreciate the opportunity to discuss critical issues surrounding the federal government's purchase of personal information [1] from businesses known as information resellers. As you are aware, the ease and speed with which people's personal information can be collected by information resellers from a wide variety of sources and made available to government and other customers has accelerated with technological advances. In recent years, security breaches at large information resellers such as ChoicePoint and LexisNexis have raised questions about how resellers and their federal customers handle people's personal information—and especially whether their practices are fully consistent with widely accepted practices for protecting the privacy and security of personal information.

Federal agency use of personal information is governed primarily by the E-Government Act of 2002 and the Privacy Act of 1974. The E-Government Act of 2002 strives to enhance protection for personal information in government information systems by requiring that agencies conduct privacy impact assessments (PIA). A PIA is an analysis of how personal information is collected, stored, shared, and managed in a federal system. The Privacy Act of 1974 [2] requires that the use of personal information be limited to predefined purposes and involve only information germane to those purposes. The provisions of the Privacy Act, in turn, are largely based on a set of principles for protecting the privacy and security of personal information, known as the Fair Information Practices, which were first proposed in 1973 by a U.S. government advisory committee [3] These principles, now widely accepted, include:

1. collection limitation,
2. data quality,
3. purpose specification,
4. use limitation,
5. security safeguards,
6. openness,
7. individual participation, and
8. Accountability [4]

These principles, with some variation, are used by organizations to address privacy considerations in their business practices and are also the basis of privacy laws and related policies in many countries, including the United States, Germany, Sweden, Australia, and New Zealand, as well as the European Union.

As agreed, my testimony today will be based primarily on the agency information contained in a report we issued in April 2006 [5]. For that report, we analyzed fiscal year

2005 contracts and other vehicles for the acquisition of personal information from information resellers by the Departments of Justice (DOJ), Homeland Security (DHS), and State (State) and the Social Security Administration (SSA). We compared relevant agency guidelines and management policies and procedures to the Fair Information Practices. We also updated the implementation status of recommendations contained in our 2006 report and analyzed provisions of the proposed Federal Agency Data Protection Act [6]. Our work was performed in accordance with generally accepted government auditing standards. Those standards require that we plan and perform the audit to obtain sufficient, appropriate evidence to provide a reasonable basis for our findings and conclusions based on our audit objectives. We believe that the evidence obtained provides a reasonable basis for our findings and conclusions based on our audit objectives.

Today, after a brief summary of the laws that govern agency use of personal information, I will summarize the information contained in our 2006 report on how the selected agencies used the personal information that they purchased from resellers and the extent to which the agencies had policies and practices that reflected the Fair Information Practices. I will also provide an update on steps taken by the agencies to address the recommendations contained in our 2006 report. Finally, I will comment on specific privacy related provisions of the proposed Federal Agency Data Protection Act.

RESULTS IN BRIEF

In fiscal year 2005, DOJ, DHS, State, and SSA reported that they planned to spend a combined total of approximately $30 million [7] to purchase personal information from resellers. The vast majority—approximately 91 percent—of the planned spending was for purposes of law enforcement (69 percent) or counterterrorism (22 percent). For example, components of DOJ (the largest user of resellers) used the information for criminal investigations, locating witnesses and fugitives, researching assets held by individuals of interest, and detecting fraud in prescription drug transactions. DHS acquired personal information to aid its immigration fraud detection and border screening programs. SSA and State purchased personal information from information resellers to detect and investigate fraud, verify identities, and determine benefits eligibility.

Agency practices for handling personal information acquired from information resellers reflected four of eight principles established by the Fair Information Practices. Agency practices generally reflected the *collection limitation*, *data quality*, *use limitation*, and *security safeguards* principles. For example, law enforcement agencies (including the Federal Bureau of Investigation and the U.S. Secret Service) generally reported that they corroborate information obtained from resellers to ensure that it is accurate when it is used as part of an investigation, reflecting the *data quality* principle that data should be accurate, current, and complete, as needed for the defined purpose. However, agencies did not always have practices for handling reseller information to fully address the *purpose specification*, *individual participation*, *openness,* and *accountability* principles. For example:

- Although agencies notified the public through *Federal Register* notices and published PIAs that they collected personal information from various sources, they did not always indicate specifically that information resellers were among those sources.
- Some agencies lacked robust audit mechanisms to ensure that use of personal information from information resellers was for permissible purposes, reflecting an uneven application of the *accountability* principle.

Contributing to agencies' uneven application of the Fair Information Practices were ambiguities in guidance from the Office of Management and Budget (OMB) on how privacy requirements apply to federal agency uses of reseller information. In addition, agencies generally lacked policies that specifically address these uses.

We made recommendations to OMB to revise privacy guidance and to the four agencies to develop specific policies for the use of personal information from resellers. The agencies generally agreed with the report and described actions initiated to address our recommendations. Since we issued our report, two of the four agencies have taken steps to address our recommendations. For example, the DHS Privacy Office incorporated specific questions in its May 2007 PIA guidance concerning use of commercial data. In addition, DOJ took steps to ensure that their system-of-records notices specifically reference their use of data from information resellers. OMB, however, has not implemented our recommendation to clarify guidance on use of commercial data.

On December 18, 2007, the Federal Agency Data Protection Act was introduced. This legislation, among other things would require that agencies (1) conduct PIAs for their uses of commercial data and (2) promulgate regulations concerning the use of commercial data brokers. We believe that these requirements are consistent with the results of our 2006 report and the recommendations we made to the agencies.

BACKGROUND

Before advanced computerized techniques, obtaining people's personal information usually required visiting courthouses or other government facilities to inspect paper-based public records, and information contained in product registrations and other business records was not generally available at all. Automation of the collection and aggregation of multiple-source data, combined with the ease and speed of its retrieval, have dramatically reduced the time and effort needed to obtain such information. Information resellers provide services based on these technological advances.

We use the term "information resellers" to refer to businesses that vary in many ways but have in common collecting and aggregating personal information from multiple sources and making it available to their customers. These businesses do not all focus exclusively on aggregating and reselling personal information. For example, Dun and Bradstreet primarily provides information on commercial enterprises for the purpose of contributing to decision making regarding those enterprises. In doing so, it may supply personal information about individuals associated with those commercial enterprises. To a certain extent, the activities of information resellers may also overlap with the functions of consumer reporting agencies, also known as credit bureaus—entities that collect and sell information about individuals' creditworthiness, among other things. To the extent that information resellers perform the functions of consumer reporting agencies, they are subject to legislation specifically addressing that industry, particularly the Fair Credit Reporting Act.

Information resellers have now amassed extensive amounts of personal information about large numbers of Americans. They supply it to customers in both government and the private sector, typically via a centralized online resource. Generally, three types of information are collected:

- *Public records* such as birth and death records, property records, motor vehicle and voter registrations, criminal records, and civil case files.
- *Publicly available information* not found in public records but nevertheless publicly available through other sources, such as telephone directories, business directories, classified ads or magazines, Internet sites, and other sources accessible by the general public.
- *Nonpublic information* derived from proprietary or nonpublic sources, such as credit header data, [8] product warranty registrations, and other application information provided to private businesses directly by consumers.

Figure 1 illustrates how these types of information are collected and aggregated into reports that are ultimately accessed by customers, including government agencies.

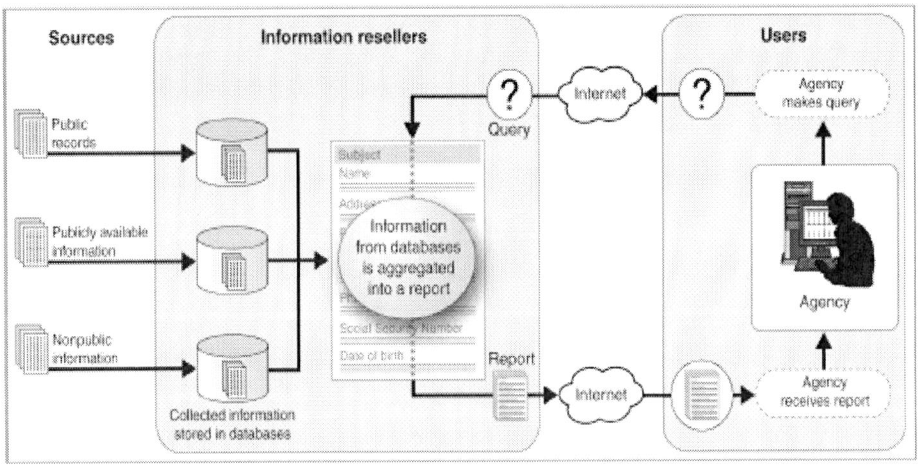

Figure 1. Typical Information Flow through Resellers to Government Customers.

Federal Laws and Guidance Govern Use of Personal Information in Federal Agencies

No single federal law governs all use or disclosure of personal information. The major requirements for the protection of personal privacy by federal agencies come from the Privacy Act of 1974 and the privacy provisions of the E-Government Act of 2002.

Federal use of personal information is governed primarily by the Privacy Act of 1974, [9] which places limitations on agencies' collection, disclosure, and use of personal information maintained in systems of records. The act describes a "record" as any item, collection, or grouping of information about an individual that is maintained by an agency and contains his or her name or another personal identifier. It also defines "system of records" as a group of records under the control of any agency from which information is retrieved by the name of the individual or by an individual identifier. The Privacy Act requires that when agencies establish or make changes to a system of records, they must notify the public by placing a notice in the *Federal Register* identifying, among other things, the type of data collected, the types of individuals about whom the information is collected, the routine uses [10] of the data, and procedures that individuals can use to review and correct their personal information. Additional provisions of the Privacy Act are discussed in the 2006 report.

The E-Government Act of 2002 requires that agencies conduct PIAs. A PIA is an analysis of how personal information is collected, stored, shared, and managed in a federal system. Under the E-Government Act and related OMB guidance, agencies must conduct PIAs (1) before developing or procuring information technology that collects, maintains, or disseminates information that is in a personally identifiable form; (2) before initiating any new data collections involving personal information that will be collected, maintained, or disseminated using information technology if the same questions are asked of 10 or more people; or (3) when a system change creates new privacy risks, for example, by changing the way in which personal information is being used.

OMB is tasked with providing guidance to agencies on how to implement the provisions of the Privacy Act and the E-Government Act and has done so, beginning with guidance on the Privacy Act, issued in 1975 [11] OMB's guidance on implementing the privacy provisions of the E-Government Act of 2002 identifies circumstances under which agencies must conduct PIAs and explains how to conduct them.

The PIA mandate in the E-Government Act of 2002 provided a mechanism by which agencies can consider privacy in the earliest stages of system development. PIAs can be an important tool to help agencies to address *openness and purpose specification* principles early in the process of developing new information systems. To the extent that PIAs are made publicly available, [12] they provide explanations to the public about such things as the information that will be collected, why it is being collected, how it is to be used, and how the system and data will be maintained and protected.

The Fair Information Practices Are Widely Agreed to Be Key Principles for Privacy Protection

The Privacy Act of 1974 is largely based on a set of internationally recognized principles for protecting the privacy and security of personal information known as the Fair Information Practices. A U.S. government advisory committee first proposed the practices in 1973 to address what it termed a poor level of protection afforded to privacy under contemporary law [13]. The Organization for Economic Cooperation and Development (OECD) [14] developed a revised version of the Fair Information Practices in 1980. This version of the principles was reaffirmed by OECD ministers in a 1998 declaration and further endorsed in a 2006 OECD report [15]. The Fair Information Practices, have, with some variation, formed the basis of privacy laws and related policies in many countries, including the United States, Germany, Sweden, Australia, and New Zealand, as well as the European Union [16].

In addition, in its 2007 report, *Engaging Privacy and Information Technology in a Digital Age*, the National Research Council [17] found that the principles of fair information practice for the protection of personal information are as relevant today as they were in 1973. Accordingly, the committee recommended that the Fair Information Practices should be extended as far as reasonably feasible to apply to private sector organizations that collect and use personal information. The eight principles of the OECD Fair Information Practices are shown in table 1.

Table 1. The OECD Fair Information Practices

Principle	Description
Collection limitation	The collection of personal information should be limited, should be obtained by lawful and fair means, and, where appropriate, with the knowledge or consent of the individual.
Data quality	Personal information should be relevant to the purpose for which it is collected, and should be accurate, complete, and current as needed for that purpose.
Purpose specification	The purposes for the collection of personal information should be disclosed before collection and upon any change to that purpose, and its use should be limited to those purposes and compatible purposes.
Use limitation	Personal information should not be disclosed or otherwise used for other than a specified purpose without consent of the individual or legal authority.
Security safeguards	Personal information should be protected with reasonable security safeguards against risks such as loss or unauthorized access, destruction, use, modification, or disclosure.
Openness	The public should be informed about privacy policies and practices, and individuals should have ready means of learning about the use of personal information.
Individual participation	Individuals should have the following rights: to know about the collection of personal information, to access that information, to request correction, and to challenge the denial of those rights.
Accountability	Individuals controlling the collection or use of personal information should be accountable for taking steps to ensure the implementation of these principles.

Source: OECD.

The Fair Information Practices are not precise legal requirements. Rather, they provide a framework of principles for balancing the need for privacy with other public policy interests, such as national security, law enforcement, and administrative efficiency. Ways to strike that balance vary among countries and according to the type of information under consideration.

Agencies Used Governmentwide Contracts to Obtain Personal Information from Information Resellers for a Variety of Purposes

DOJ, DHS, State, and SSA reported approximately $30 million through contracts with information resellers in fiscal year 2005 [18]. The agencies reported using personal information obtained from resellers for a variety of purposes including law enforcement, counterterrorism, fraud detection/prevention, and debt collection. In all, approximately 91 percent of agency uses of reseller data were in the categories of law enforcement (69 percent) or counterterrorism (22 percent). Figure 2 details contract values categorized by their reported use.

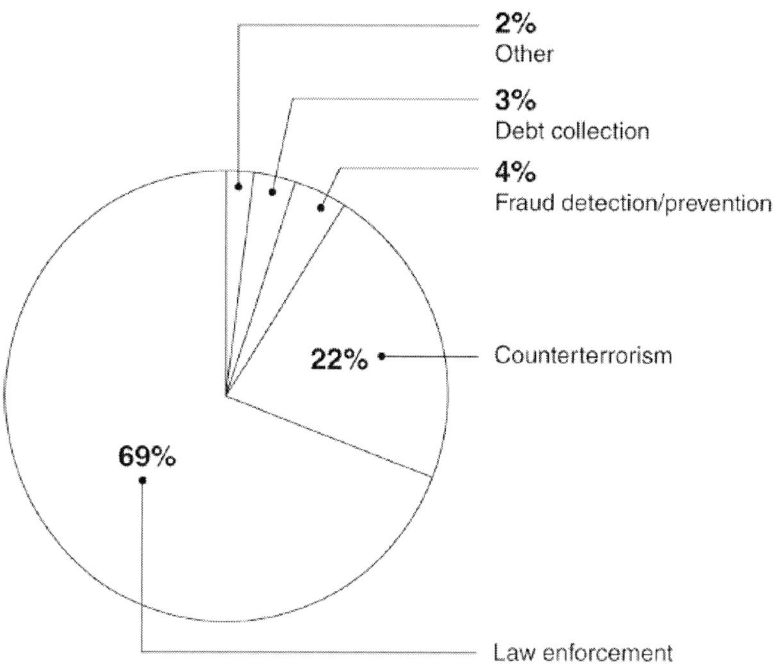

Figure 2. Fiscal Year 2005 Contractual Vehicles Enabling the Use of Personal Information from Information Resellers, Categorized by Reported Use.

DOJ, which accounted for about 63 percent of the funding, mostly used the data for law enforcement and counterterrorism. DHS also used reseller information primarily for law enforcement and counterterrorism. State and SSA reported acquiring personal information from information resellers for fraud prevention and detection, identity verification, and benefits eligibility determination.

DOJ and DHS Used Information Resellers Primarily for Law Enforcement and Counterterrorism

In fiscal year 2005, DOJ and its components reported approximately $19 million through contracts with a wide variety of information resellers, primarily for purposes related to law enforcement (75 percent) and counterterrorism (18 percent). The Federal Bureau of Investigation (FBI), which is DOJ's largest user of information resellers, used reseller information to, among other things, analyze intelligence and detect terrorist activities in support of ongoing investigations by law enforcement agencies and the intelligence community. In this capacity, resellers provided the FBI's Foreign Terrorist Tracking Task Force with names, addresses, telephone numbers, and other biographical and demographical information as well as legal briefs, vehicle and boat registrations, and business ownership records [19].

The Drug Enforcement Administration (DEA), the second largest DOJ user of information resellers in fiscal year 2005, obtained reseller data primarily to detect fraud in prescription drug transactions.20 Agents used reseller data to detect irregular prescription patterns for specific drugs and trace this information to the pharmacy and prescribing doctor [21].

DHS and its components reported that they used information reseller data in fiscal year 2005 primarily for law enforcement purposes, such as developing leads on subjects in criminal investigations and detecting fraud in immigration benefit applications (part of enforcing immigration laws). DHS's largest investigative component, the U.S. Immigration and Customs Enforcement, is also its largest user of personal information from resellers. It collected data such as address and vehicle information for criminal investigations and background security checks. Another DHS component, U.S. Customs and Border Protection, conducts queries on people, businesses, property. The Federal Emergency Management Agency, an additional component, used an information reseller to detect fraud in disaster assistance applications.

DHS also reported using information resellers in its counterterrorism efforts. For example, the Transportation Security Administration (TSA), a DHS component, used data obtained from information resellers as part of a test associated with the development of its domestic passenger prescreening program, called Secure Flight [22]. TSA planned for Secure Flight to compare domestic flight reservation information submitted to TSA by aircraft operators with federal watch lists of individuals known or suspected of activities related to terrorism [23].

SSA and State Used Information Resellers Primarily for Fraud Prevention and Detection

In an effort to ensure the accuracy of Social Security benefit payments, the SSA and its components reported approximately $1.3 million in contracts with information resellers in fiscal year 2005 for purposes relating to fraud prevention (such as skiptracing), [24] confirming suspected fraud related to workers' compensation payments, obtaining information on criminal suspects for follow-up investigations, and collecting debts. For example, the Office of the Inspector General (OIG), the largest user of information reseller data at SSA, used several information resellers to assist investigative agents in detecting benefits abuse by Social Security claimants and to assist agents in locating claimants. Regional office agents may also use reseller data in investigating persons suspected of claiming disability fraudulently.

State and its components reported approximately $569,000 in contracts with information resellers for fiscal year 2005, mainly to support investigations of passport-related activities. For example, several components accessed personal information to validate familial relationships, birth and identity data, and other information submitted on immigrant and nonimmigrant visa petitions. State also used reseller data to investigate passport and visa fraud cases.

Agencies Lacked Policies on Use of Reseller Data, and Practices Do Not Consistently Reflect the Fair Information Practices

Agencies generally lacked policies that specifically addressed their use of personal information from commercial sources (although DHS Privacy Office officials reported in 2006 that they were drafting such a policy [25]), and agency practices for handling personal information acquired from information resellers did not always fully reflect the Fair Information Practices. Specifically, agency practices generally reflected four of the eight Fair Information Practices.

As table 2 shows, the *collection limitation*, *data quality*, *use limitation*, and *security safeguards* principles were generally reflected in agency practices. For example, several agency components (specifically, law enforcement agencies such as the FBI and the U.S. Secret Service) reported that in practice, they generally corroborate information obtained from resellers when it is used as part of an investigation. This practice is consistent with the principle of *data quality*.

Agency policies and practices with regard to the other four principles were uneven. Specifically, agencies did not always have policies or practices in place to address the *purpose specification*, *openness,* and *individual participation* principles with respect to reseller data. The inconsistencies in applying these principles as well as the lack of specific agency policies can be attributed in part to ambiguities in OMB guidance regarding the applicability of the Privacy Act to information obtained from resellers. Further, privacy impact assessments, a valuable tool that could address important aspects of the Fair Information Practices, were often not conducted.

Finally, components within each of the four agencies did not consistently hold staff accountable by monitoring usage of personal information from information resellers and ensuring that it was appropriate; thus, their application of the fourth principle, *accountability,* was uneven.

Agency procedures generally reflected the *collection limitation*, *data quality*, *use limitation*, and *security safeguards* principles. Regarding collection limitation, for most law-enforcement and counterterrorism purposes (which accounted for 90 percent of usage in fiscal year 2005), agencies generally limited their personal data collection in that they reported obtaining information only on specific individuals under investigation or associates of those individuals. Regarding *data quality*, agencies reported taking steps to mitigate the risk of inaccurate information reseller data by corroborating information obtained from resellers. Agency officials described the practice of corroborating information as a standard element of conducting investigations.

Likewise, for non-law-enforcement use, such as debt collection and fraud detection and prevention, agency components reported that they mitigated potential problems with the accuracy of data provided by resellers by obtaining additional information from other sources when necessary. As for *use limitation*, agency officials said their use of reseller information was limited to distinct purposes that were generally related to law enforcement or counterterrorism. Finally, while we did not assess the effectiveness of information security at any of these agencies, we found that all four had measures in place intended to safeguard the security of personal information obtained from resellers [26].

Limitations in the Applicability of the Privacy Act and Ambiguities in OMB Guidance Contributed to an Uneven Adherence to the *Purpose Specification*, *Openness*, and *Individual Participation* Principles

The *purpose specification, openness*, and *individual participation* principles stipulate that individuals should be made aware of the purpose and intended uses of the personal information being collected about them, and, if necessary, have the ability to access and correct their information. These principles are reflected in the Privacy Act requirement for agencies to publish in the *Federal Register*, "upon establishment or revision, a notice of the existence and character of a system of records." This notice is to include, among other things, the categories of records in the system as well as the categories of sources of records [27].

In a number of cases, agencies using reseller information did not adhere to the *purpose specification* or *openness* principles in that they did not notify the public that they were using such information and did not specify the purpose for their data collections. Agency officials said that they generally did not prepare system-of-records notices that would address these principles because they were not required to do so by the Privacy Act. The act's vehicle for public notification—the system-of-records notice—is required of an agency only when the agency collects, maintains, and retrieves personal data in the way defined by the act or when a contractor does the same thing explicitly on behalf of the government. Agencies generally did not issue system-of-records notices specifically for their use of information resellers largely because information reseller databases were not considered "systems of records operated by or on behalf of a government agency" and thus were not considered subject to the provisions of the Privacy Act [28]. OMB guidance on implementing the Privacy Act does not specifically refer to the use of reseller data or how it should be treated. According to OMB and other agency officials, information resellers operate their databases for multiple customers, and federal agency use of these databases does not amount to the operation of a system of records on behalf of the government. Further, agency officials stated that merely querying information reseller databases did not amount to agency "maintenance" of the personal information being queried and thus also did not trigger the provisions of the Privacy Act. In many cases, agency officials considered their use of resellers to be of this type—essentially "ad hoc" querying or "pinging" of reseller databases for personal information about specific individuals, which they believed they were not doing in connection with a formal system of records.

Table 2. Application of Fair Information Practices to the Reported Handling of Personal Information from Data Resellers at Four Agencies

Principle	Agency application of principle	Agency practices
Collection limitation. The collection of personal information should be limited, should be obtained by lawful and fair means, and, where appropriate, with the knowledge or consent of the individual.	General	Agencies limited personal data collection to individuals under investigation or their associates.
Data quality. Personal information should be relevant to the purpose for which it is collected, and should be accurate, complete, and current as needed for that purpose.	General	Agencies corroborated information from resellers and did not take actions based exclusively on such information.
Purpose specification. The purpose for the collection of personal information should be disclosed before collection and upon any change to that purpose, and its use should be limited to that purpose and compatible purposes.	Uneven	Agency system-of-records notices did not generally reveal that agency systems could incorporate information from data resellers. Agencies also generally did not conduct privacy impact assessments for their systems or programs that involve use of reseller data.
Use limitation. Personal information should not be disclosed or otherwise used for other than a specified purpose without consent of the individual or legal authority.	General	Agencies generally limited their use of personal information to specific investigations (including law enforcement, counterterrorism, fraud detection, and debt collection).
Security safeguards. Personal information should be protected with reasonable security safeguards against risks such as loss or unauthorized access, destruction, use, modification, or disclosure.	General	Agencies had security safeguards such as requiring passwords to access databases, basing access rights on need to know, and logging search activities (including "cloaked logging," which prevents the vendor from monitoring search content).
Openness. The public should be informed about privacy policies and practices, and individuals should have ready means of learning about the use of personal information.	Uneven	See *Purpose specification* above. Agencies did not have established policies specifically addressing the use of personal information obtained from resellers.
Individual participation. Individuals should have the following rights: to know about the collection of personal information, to access that information, to request correction, and to challenge the denial of those rights.	Uneven	See *Purpose specification* above. Because agencies generally did not disclose their collections of personal information from resellers, individuals were often unable to exercise these rights.
Accountability. Individuals controlling the collection or use of personal information should be accountable for taking steps to ensure the implementation of these principles.	Uneven	Agencies did not generally monitor usage of personal information from information resellers to hold users accountable for appropriate use; instead, they relied on users to be responsible for their behavior. For example, agencies may instruct users in their responsibilities to use personal information appropriately, have them sign statements of responsibility, and have them indicate what permissible purpose a given search fulfills.

Legend:

General = policies or procedures to address all major aspects of a particular principle.

Uneven = policies or procedures addressed some, but not all, aspects of a particular principle or some but not all agencies and components had policies or practices in place addressing the principle.

Note: We did not independently assess the effectiveness of agency information security programs. Our assessment of overall agency application of the Fair Information Practices was based on the policies and management practices described by the Department of State and SSA as a whole and by major components of DOJ and DHS. We did not obtain information on smaller components of DOJ and DHS.

In other cases, however, agencies maintained information reseller data in systems for which system-of-records notices had been previously published. For example, law enforcement agency officials stated that, to the extent they retain the results of reseller data queries, this collection and use is covered by the system-of-records notices for their case file systems. However, in preparing such notices, agencies generally did not specify that they were obtaining information from resellers. Among system-of-records notices that were identified by agency officials as applying to the use of reseller data, only one—TSA's system-of-records notice for the test phase of its Secure Flight program—specifically identified the use of information reseller data [29].

In several of these cases, agency sources for personal information were described only in vague terms, such as "private organizations," "other public sources," or "public source material," when information was being obtained from information resellers.

The inconsistency with which agencies specify resellers as a source of information in system-of-records notices is due in part to ambiguity in OMB guidance, which states that "for systems of records which contain information obtained from sources other than the individual to whom the records pertain, the notice should list the types of sources used."[30] Although the guidance is unclear as to what would constitute adequate disclosure of "types of sources," OMB and DHS Privacy Office officials agreed that to the extent that reseller data is subject to the Privacy Act, agencies should specifically identify information resellers as a source and that merely citing public records information does not sufficiently describe the source.

Aside from certain law enforcement exemptions [31] to the Privacy Act, adherence to the *purpose specification* and *openness* principles is critical to preserving a measure of individual control over the use of personal information. Without clear guidance from OMB or specific policies in place, agencies have not consistently reflected these principles in their collection and use of reseller information. As a result, without being notified of the existence of an agency's information collection activities, individuals have no ability to know that their personal information could be obtained from commercial sources and potentially used as a basis, or partial basis, for taking action that could have consequences for their welfare.

Privacy Impact Assessments Could Address Openness and Purpose Specification Principles but often Were not Conducted

PIAs can be an important tool to help agencies to address *openness and purpose specification* principles early in the process of developing new information systems. To the extent that PIAs are made publicly available, [32] they provide explanations to the public about things such as the information that will be collected, why it is being collected, how it is to be used, and how the system and data will be maintained and protected.

However, few agency components reported developing PIAs for their systems or programs that make use of information reseller data. As with system-of-records notices, agencies often did not conduct PIAs because officials did not believe they were required. Current OMB guidance on conducting PIAs is not always clear about when they should be conducted. According to guidance from OMB, a PIA is required by the E-Government Act when agencies "systematically incorporate into existing information systems databases of information in identifiable form purchased or obtained from commercial or public sources" [33]. However, the same guidance also instructs agencies that "merely querying a database on an ad hoc basis does not trigger the PIA requirement." Reported uses of reseller data were generally not described as a "systematic" incorporation of data into existing information systems; rather, most involved querying a database and, in some cases, retaining the results of these queries. OMB officials stated that agencies would need to make their own judgments on whether retaining the results of searches of information reseller databases constituted a "systematic incorporation" of information.

Until PIAs are conducted more thoroughly and consistently, the public is likely to remain incompletely informed about agency purposes and uses for obtaining reseller information.

Agencies Often Did Not Have Practices in Place to Ensure Accountability for Proper Handling of Information Reseller Data

According to the *accountability* principle, individuals controlling the collection or use of personal information should be accountable for ensuring the implementation of the Fair Information Practices. This means that agencies should take steps to ensure that they use personal information from information resellers appropriately.

Agencies described using activities to oversee their use of reseller information that were largely based on trust in the individual user to use the information appropriately, rather than on management oversight of usage details. For example, in describing controls placed on the use of commercial data, officials from component agencies identified measures such as instructing users that reseller data are for official use only and requiring users to sign statements attesting 1) to their need to access information reseller databases and 2) that their use will be limited to official business. Additionally, agency officials reported that their users are required to select from a list of vendor-defined "permissible purposes" (for example, law enforcement, transactions authorized by the consumer) before conducting a search on reseller databases.

While these practices appear consistent with the accountability principle, they are focused on individual user responsibility instead of monitoring and oversight. Agencies did not have practices in place to obtain reports from resellers that would allow them to monitor usage of reseller databases at a detailed level. Although agencies generally receive usage reports from the information resellers, these reports are designed primarily for monitoring costs. Further, these reports generally contained only high-level statistics on the number of searches and databases accessed, not the contents of what was actually searched, thus limiting their utility in monitoring usage.

To the extent that federal agencies do not implement methods such as user monitoring or auditing of usage records, they provide limited accountability for their usage of information reseller data and have limited assurance that the information is being used appropriately.

Not All Agencies Have Taken Steps to Address Our Recommendations

In our report, we recommended that the agencies develop specific policies for the collection, maintenance, and use of personal information obtained from resellers. We also recommended that OMB revise its privacy guidance to clarify the applicability of requirements for public notices and privacy impact assessments to agency use of personal information from resellers and direct agencies to review their uses of such information to ensure it is explicitly referenced in privacy notices and assessments. The agencies generally agreed with our findings and described actions initiated to address our recommendations.

Since the issuance of our 2006 report, two of the four agencies have taken action to address our recommendation. For example, the DHS Privacy Office incorporated specific questions in its May 2007 PIA guidance concerning use of commercial data. The guidance requires programs that use commercial or publicly available data to explain why and how such data are used. Further, the guidance for systems that use or rely on commercial data requires an explanation of how data accuracy and integrity are preserved and the reliability of the data assessed with regard to its value to the purpose of the system. According to DHS Privacy Office officials, after identifying use of commercial data through the PIA process, the

Privacy Office works with the relevant DHS component to review uses of commercial data to ensure appropriate controls are in place and that the planned uses are appropriately disclosed in privacy notices. In addition, officials at DOJ informed us that the Privacy and Civil Liberties Office has in place a verbal agreement with agency components that there are to be no bulk acquisitions of commercial data and that when the agency takes in data from commercial sources, there should be a valid system-of-records notice that specifically identifies commercial data as a source. Further, DOJ has updated several of its system-of-records notices to reflect their use of data from information resellers. SSA and State have not yet addressed our recommendation.

However, OMB has not addressed our recommendations. In an August 2006 letter to congressional committees in response to the recommendations contained in our April 2006 report, OMB noted that work on the protection of personal information through the Identity Theft Task Force was ongoing and that following the completion of this work, they would consider issuing appropriate clarifying guidance concerning reseller data. Since then, OMB's efforts on the Identity Theft Task Force have been completed and on May 22, 2007 OMB issued M-07-16, "Safeguarding Against the Breach of Personally Identifiable Information." To date, OMB has not issued additional clarifying guidance concerning reseller data.

Privacy Provisions of the Proposed Federal Agency Data Protection Act Are Consistent with Our Recommendations

The Federal Agency Data Protection Act was introduced on December 18, 2007. Among other things, the legislation contains privacy provisions that would require agencies to conduct PIAs when "purchasing or subscribing for a fee to information in identifiable form from a data broker." We believe that such a requirement is consistent with the recommendations contained in our report, particularly given the debate concerning whether or not agencies "systematically incorporate" information or are "merely pinging or querying the information." Our report found that PIAs could serve to address certain Fair Information Practice principles such as *purpose specification* and *openness*, but often were not conducted. Such a requirement could more readily ensure agencies perform these assessments. Further, since OMB has not clarified its guidance on this issue, a requirement in law could provide needed direction to agencies.

The proposed Federal Agency Data Protection Act would also require each agency to prescribe regulations that specify, among other things, the personnel permitted to access, analyze, or otherwise use commercial reseller databases. This legislation is consistent with our recommendation that agencies develop policies concerning their use of personal information from information resellers.

In summary, services provided by information resellers are important to federal agency functions such as law enforcement and fraud protection and identification. While agencies have taken steps to adhere to some Fair Information Practices such as the *collection limitation*, *data quality*, *use limitation*, and *security safeguards* principles, they have not taken all the steps they could to reflect others—or to use the specific processes of the Privacy Act and E-Government Act requirements—in their handling of reseller data. Because OMB privacy guidance does not clearly address information reseller data, agencies are left largely on their own to determine how to satisfy legal requirements and protect privacy when acquiring and using reseller data. Since we issued our report in 2006, two of the four agencies have taken steps to address our recommendations. However, OMB has not modified its guidance. Without current and specific guidance, the government risks continued uneven adherence to important, well-established privacy principles and lacks assurance that the privacy rights of individuals are being adequately protected. Absent action from OMB to revise guidance, privacy provisions contained in the proposed Federal Agency Data Protection Act could clarify the need to conduct privacy impact assessments wherever reseller data are involved and promote the development of agency policies and procedures concerning the use of such data. We believe these provisions are consistent with the results and recommendations contained in our 2006 report.

Mr. Chairman, this concludes my testimony today. I would be happy to answer any questions you or other members of the subcommittee may have.

REFERENCES

[1] For purposes of this chapter, the term personal information is defined as any information about an individual maintained by an agency, including (1) any information that can be used to distinguish or trace an individual's identity, such as name, Social Security number, date and place of birth, mother's maiden name, or biometric records, and (2) any other information that is linked or linkable to an individual, such as medical, educational, financial, and employment information.

[2] The Privacy Act of 1974, Pub. L. No. 93-579, 88 Stat. 1896 (codified as amended at 5 U.S.C. § 552a) provides safeguards against an invasion of privacy through the misuse of records by federal agencies and allows citizens to learn how their personal information is collected, maintained, used, and disseminated by the federal government.

[3] Congress used the committee's final report as a basis for crafting the Privacy Act of 1974. See U.S. Department of Health, Education, and Welfare, *Records, Computers, and the Rights of Citizens: Report of the Secretary's Advisory Committee on Automated Personal Data Systems* (Washington, D.C.; July 1973).

[4] Descriptions of these principles are shown in table 1.

[5] GAO, *Personal Information: Agency and Reseller Adherence to Key Privacy Principles*, GAO-06-421 (Washington, D.C.: Apr. 4, 2006).

[6] H.R. 4791, Federal Agency Data Protection Act, 110th Cong., introduced by Representative Wm. Lacy Clay, December 18, 2007.

[7] This figure may include uses that do not involve personal information. Except for instances where the reported use was primarily for legal research, agency officials were unable to separate the dollar values associated with use of personal information from uses for other purposes (for example, LexisNexis and West provide news and legal research in addition to public records). The four agencies obtained personal information from resellers primarily through two general-purpose governmentwide contract vehicles—the Federal Supply Schedule of the General Services Administration and the Library of Congress's Federal Library and Information Network.

[8] Credit header data are the nonfinancial identifying information located at the top of a credit report, such as name, current and prior addresses, telephone number, and Social Security number.

[9] The Privacy Act of 1974, Pub. L. No. 93-579, 88 Stat. 1896 (codified as amended at 5 U.S.C. § 552a) provides safeguards against an invasion of privacy through the misuse of records by federal agencies and allows citizens to learn how their personal information is collected, maintained, used, and disseminated by the federal government.

[10] Under the Privacy Act of 1974, the term "routine use" means (with respect to the disclosure of a record) the use of such a record for a purpose that is compatible with the purpose for which it was collected. 5 U.S.C. § 552a (a(7)).

[11] OMB, "Privacy Act Implementation: Guidelines and Responsibilities," *Federal Register*, Volume 40, Number 132, Part III, pages 28948-28978 (Washington, D.C.; July 9, 1975). Since the initial Privacy Act guidance of 1975, OMB has periodically published additional guidance. Further information regarding OMB Privacy Act guidance can be found on the OMB Web site at http://www.whitehouse.gov/omb/inforeg/infopoltech.html.

[12] The E-Government Act requires agencies, if practicable, to make PIAs publicly available through agency Web sites, publication in the *Federal Register* or by other means. Pub. L. No. 107-347, § 208 (b)(1)(B)(iii).

[13] U.S. Department of Health, Education, and Welfare, *Records, Computers and the Rights of Citizens*.

[14] OECD, *Guidelines on the Protection of Privacy and Transborder Flow of Personal Data* (Sept. 23, 1980). The OECD plays a prominent role in fostering good governance in the public service and in corporate activity among its 30 member countries. It produces internationally agreed-upon instruments, decisions, and recommendations to promote rules in areas where multilateral agreement is necessary for individual countries to make progress in the global economy.

[15] OECD, *Making Privacy Notices Simple: An OECD Report and Recommendations* (July 24, 2006).

[16] European Union Data Protection Directive ("Directive 95/46/EC of the European Parliament and of the Council of 24 October 1995 on the Protection of Individuals with Regard to the Processing of Personal Data and the Free Movement of Such Data") (1995).

[17] National Research Council of the National Academies, *Engaging Privacy and Information Technology in a Digital Age* (Washington, D.C.; 2007).

[18] This figure comprises contracts and task orders with information resellers that included the acquisition and use of personal information. However, some of these funds may have been for uses that do not involve personal information; we could not omit all such uses because agency officials were not always able to separate the amounts associated with the use of personal information from those for other uses (for example, LexisNexis and West provide news and legal research in addition to public records). In some instances, where the reported use was primarily for legal research, we omitted these funds from the total.

[19] GAO, *Data Mining: Agencies Have Taken Key Steps to Protect Privacy in Selected Efforts, but Significant Compliance Issues Remain,* GAO-05-866 (Washington, D.C.: Aug. 15, 2005).

[20] DEA's mission includes enforcing laws pertaining to the manufacture, distribution, and dispensing of legally produced controlled substances.

[21] The personal information contained in this information reseller database is limited to the prescribing doctor and does not contain personal patient information.

[22] For an assessment of privacy issues associated with the Secure Flight commercial data test, see GAO, *Aviation Security: Transportation Security Administration Did Not Fully Disclose Uses of Personal Information during Secure Flight Program Testing in Initial Privacy Notices, but Has Recently Taken Steps to More Fully Inform the Public* ,GAO-05-864R (Washington, D.C.: July 22, 2005).

[23] TSA's current plans for Secure Flight do not include the use of reseller information.

[24] Skiptracing is the process of locating people who have fled in order to avoid paying debts.

[25] Subsequent to the 2006 report, the DHS Privacy Office took steps to develop guidance on the use of personal information from information resellers in its PIA guidance.

[26] Although we did not assess the effectiveness of information security at any agency as part of this review, we have previously reported on weaknesses in almost all areas of information security controls at 24 major agencies, including DOJ, DHS, State, and SSA. For additional information see GAO, *Information Security: Weaknesses Persist at Federal Agencies Despite Progress Made in Implementing Related Statutory Requirements* ,GAO-05-552 (Washington, D.C.: July 15, 2005) and *Information Security: Department of Homeland Security Needs to Fully Implement Its Security Program*, GAO-05-700 (Washington, D.C.: June 17, 2005).

[27] 5 U.S.C. § 552a(e)(4)(C) and (I). The Privacy Act allows agencies to claim an exemption from identifying the categories of sources of records for records compiled for criminal law enforcement purposes, as well as for a broader category of uses, including investigative records compiled for criminal or civil law enforcement purposes.

[28] The act provides for its requirements to apply to government contractors when agencies contract for the operation by or on behalf of the agency, a system of records to accomplish an agency function. 5 U.S.C. § 552a(m).

[29] As we have previously reported, this notice did not fully disclose the scope of the use of reseller data during the test phase. See GAO, *Aviation Security: Transportation Security Administration Did Not Fully Disclose Uses of Personal Information during Secure Flight Program Testing in Initial Privacy Notices, but Has Recently Taken Steps to More Fully Inform the Public*, GAO-05-864R (Washington, D.C.: July 22, 2005).

[30] OMB, "Privacy Act Implementation: Guidelines and Responsibilities," *Federal Register,* Volume 40, Number 132, Part III, p. 28964 (Washington, D.C.: July 9, 1975).

[31] The Privacy Act allows agencies to claim exemptions if the records are used for certain purposes. 5 U.S.C. § 552a (j) and (k). For example, records compiled for criminal law enforcement purposes can be exempt from the access and correction provisions. In general, the exemptions for law enforcement purposes are intended to prevent the disclosure of information collected as part of an ongoing investigation that could impair the investigation or allow those under investigation to change their behavior or take other actions to escape prosecution. In most cases where officials identified system-of-record notices associated with reseller data collection for law enforcement purposes, agencies claimed this exemption.

[32] The E-Government Act requires agencies, if practicable, to make privacy impact assessments publicly available through agency Web sites, publication in the *Federal Register*, or by other means. Pub. L. No. 107-347, § 208 (b)(1)(B)(iii).

[33] OMB, *Guidance for Implementing the Privacy Provisions of the E-Government Act of 2002*, Memorandum M-03-22 (Washington, D.C.: Sept. 26, 2003).

In: Computer Networks, Policies and Applications
Editors: C. M. Berger and A. D. Streusse
ISBN: 978-1-61209-088-7

Chapter 10

INTERNET AND COMPUTER-MEDIATED TECHNOLOGIES AS NEW TOOLS IN THE STUDY OF SHYNESS: A REVIEW OF RECENT SURVEY- AND LABORATORY-BASED STUDIES

Paul M. Brunet and Louis A. Schmidt*
Department of Psychology, Neuroscience and Behaviour
McMaster University, Ontairio, Canada

ABSTRACT

Internet and computer technologies allow for the investigation of traditional questions in personality and experimental social psychology regarding person-context interactions in new ways. In this paper, we review recent survey- and laboratory-based studies that have used the Internet and computer technologies to study shyness. We also discuss how these technologies can be used as laboratory tools to manipulate context in the study of shyness and social behavior. We conclude with suggestions for future work using computer technologies in both basic and applied areas of research on shyness and social anxiety.

INTRODUCTION

Shyness is an anxious pre-occupation of the self in response to real or anticipation of social interaction. Over 90% of the population have episodes of shyness at some point in their lives (Zimbardo, 1977). These infrequent incidences are usually the result of a contextual cue such as another person, a setting, or an event. For example, an adolescent might feel shy on her first date, or a new employee might feel shy meeting the CEO of the company. Although these infrequent moments of discomfort during a social situation are temporary and may have

* Please address all correspondence to Louis A. Schmidt, Department of Psychology, Neuroscience and Behaviour, McMaster University, Hamilton, Ontario, Canada, L8S 4K1; email: schmidtl@mcmaster.ca

no detrimental effects on the individual's well-being, this scenario is not true for every person. A subset of the population (10-15%) can be characterized as extremely shy or temperamentally shy (Kagan, 1994). For such individuals, shyness is an enduring personality trait which is stable both across time and across different domains of life (e.g., work place, dating, and school). It is important to note that despite the stability of shyness in some individuals, their shyness is influenced by context.

Recent advances in computer technologies such as the Internet allow for the investigation of old questions in personality and experimental social psychology regarding person-context interactions in new ways. In this paper, we review research that has led to our current understanding of shyness, and how that understanding can be enhanced using new technology such as the Internet and computer-mediated communication to address traditional questions related to person-context interactions. Although the extant literature has often used the term shyness interchangeably with social anxiety and social phobia, the work described herein is focused largely on shyness. Where related constructs such as social anxiety and social phobia are used, this distinction is clearly noted.

The paper is divided among three parts. The first part focuses on the influence of shyness on studies involving traditional face-to-face (FTF) social communicative behavior. Part two addresses how today's technology (i.e., Internet and computer-mediated technologies) can be used as tools to study traditional questions in personality and experimental social psychology in news ways, and how the technology can be used to study shyness and context interactions on social communicative behaviour in particular. The third part provides suggestions for future directions in using the Internet as an intervention and treatment tool for extreme shyness, social anxiety, and social phobia.

Shyness and Social Communicative Behaviour

Considering that shyness is defined by feelings of heightened self-consciousness and self-preoccupation in social settings (Crozier, 1979; Fatis, 1983), it is not surprising that over the last few decades much research has been dedicated to investigating the influence of shyness on social communicative behaviour. Through self-reported behaviour studies or experimentally designed observations of behaviour, we have a good understanding of how shy individuals behave socially in FTF settings.

One of the earliest studies in the area of shyness and FTF social communication was conducted by Pilkonis (1977). Participants were observed in an opposite-sex interaction, an interaction with the experimenter, and in the preparation and delivery of a speech. Pilkonis (1977) found that shy individuals, when interacting with an opposite-sex confederate, spoke less frequently, took longer to initiate a conversation, spoke for a smaller percentage of the time, and were less likely to break any lulls in conversation compared to their non-shy counterparts. However, despite making fewer utterances, the length of the shy and non-shy individuals' utterances was equivalent. Pilkonis (1977) concluded that shy individuals engage in reactive rather than initiative conversational styles. These differences were more apparent with males than females. Shy males made fewer glances, and less eye contact, and spent a shorter percentage of time looking at the confederate than non-shy males; on the other hand, shy females displayed more nodding and smiling than non-shy females or males in general. In the speech delivery setting, shyness was not related to poorer performance, despite the shy individuals reporting higher anxiety. This finding implies that shyness influences verbal behaviour in social settings, but not in speaking situations in general.

Thompson and Rapee (2002) reported similar findings by observing participants in an unstructured dyadic conversation (i.e., they were given no instructions to converse) and in a structured conversation (i.e., participants were instructed to get to know one another) with a confederate. Socially anxious participants differed from their non-socially anxious counterparts only in the unstructured conversation. Their overall molecular behaviour scores (e.g., conversation skills, voice quality) were worse compared to the non-socially anxious group in the unstructured conversation, but not in the structured conversations. Therefore, just as reported by Pilkonis (1977), being assigned a task seems to help shy individuals with their speaking and conversational skills. Unfortunately, outside of a controlled laboratory setting, shy individuals will be faced with many more situations of unstructured than structured conversations.

Another study, which refuted and supported Pilkonis'(1977) findings, investigated the influence on shyness on mixed-sex dyads who were asked to get to know one another (Garcia, Stinson, Ickes, Bissonnette, and Briggs, 1991). Shyness was related to lower frequency and duration of the dyad's speaking turns (i.e., the combined score of both participants) and the number of questions asked during the conversation, but only for the male shyness. The lower frequency of turns matched Pilkonis' (1977) finding, but the shorter duration of turns did not. However, the difference may be due to type of measurement. Pilkonis (1977) used number of words per utterance as the measure, whereas it appears that Garcia and colleagues (1991) used the amount of time that utterances took. It is possible that shy males use the same number of words, but take less time to speak than other individuals. Male shyness was also related to non-verbal behaviour such as fewer smiles, less laughter, and fewer mutual gazes by the dyad. Interestingly, male shyness not only influenced the actor's behaviour, it also negatively influenced their partner's behaviour. Female shyness did not influence the actor's behaviours or the partner's behaviour.

Manning and Ray's (1993) dyadic study, in which participants conversed for 10 minutes, also supported Pilkonis' (1977) findings. Shy individuals do not seek speakership or attempt to end silences in conversation as frequently as non-shy individuals. Manning and Ray (1993) also observed that shy individuals spent more time discussing the setting (i.e., the room in which they are in) and attempt to maintain a favoured topic (i.e., stick to one preferred topic). However, it is important to note that the authors did not present any statistical analyses to support their observations.

In a survey study, Hill (1989) asked participants about the behaviours they exhibit in five different settings: when conversing with the opposite sex, when in class, when at a party, when in a checkout line, and when being introduced to a new person. Shy individuals reported initiating fewer conversations, asking fewer questions, and engaging in less small talk in all five settings. Interestingly, their reported opinions as to what constitutes appropriate behaviour matched the opinions given by non-shy participants, and matched the behaviour exhibited by non-shy individuals, but not their own. The implication is that shy individuals do recognize what is considered appropriate behaviour, but do not manage to exhibit that behaviour.

Research also demonstrated that shy individuals are less likely to self-disclose information about themselves during conversation. In a survey study, Matsushima, Shiomi, and Kuhlman (2000) found that participants' (N= 433) self-reported shyness was predictive of weaker social skills and self-disclosures. The influence of shyness on self-disclosure was moderated by social skills. Matsushima et al. (2000) suggested that increases in shyness were related to decreases in social skills and active interpersonal relationship skills, which then leads to less self-disclosure behaviour. Similarly, in a dyadic study, socially anxious participants paired with a confederate assigned to a high intimacy condition (i.e., confederate made highly personal disclosures) provided less intimate disclosures than non-socially anxious participants (Melesko and Alden, 1993). However, in a low intimacy condition (i.e., confederate made superficial disclosures), socially anxious individuals provide more personal disclosures than their peers. Overall, both groups disclosed more intimately in the high intimate condition than in the low intimate condition. Taken together, the findings of the survey study (Matsushima et al., 2000) and of the dyadic study (Melesko and Alden, 1993) demonstrated that shy and socially anxious individuals disclose less information and less intimate information about themselves in most face-to-face situations.

The pattern of verbal communication in people who are shy also appears to generalize to written expression. By coding writing samples of 2406 undergraduate students, Pennebaker and King (1999) found that introversion (a construct related to shyness) was moderately related to a higher use of tentative words (e.g., perhaps), and exclusive words (e.g., without), negations (e.g., not), and related to lower use of inclusion words (e.g., with). Additionally, they found that introversion was significantly related to a decrease in positive emotion word use (i.e., happy, beautiful) and to an increase in negative emotion word use (e.g., ugly, sadness).

The Use of the Internet and Computer Technologies as New Tools to Study Shyness: Recent Survey and Laboratory-Based Studies

Over the last two decades, the Internet has developed from a means for researchers to share information to a mainstream technological resource with numerous functions. The Internet is available in most work, school, and home settings in developed countries. Computer developers are always upgrading and updating the associated features of the Internet to facilitate its users' ability to navigate the Internet to satisfy their needs. The increase of accessibility and user friendliness has lead to the Internet becoming one of the leading social media available.

Through chat rooms, message boards, blogs, and instant messengers (e.g., Yahoo Messenger or Windows Live Messenger), Internet users have a quick and easy method of maintaining current relationships with friends and family and initiating new relationships with individuals located anywhere in the world. By becoming a member of a social network website (e.g., Facebook), individuals are able to upload pictures, share their interests, update their status (e.g., location, job, marital status), and send messages to friends. Consequently, the Internet has not only mimicked previous social media (e.g., the telephone), it has created opportunities for new types of socializing specific to itself, which have been embraced by its users. Undoubtedly, this new social medium has piqued the interest of personality and health psychologists and professionals from various fields (e.g., business, communications, and sociology; see Taylor, 2002, for a Special Issue devoted to the topic).

Two lines of research based on the Internet have been developed: the first to determine who and for what reasons the Internet is used with survey studies (Amichai-Hamburger, Wainapel, and Fox, 2002; Chak and Leung, 2004; Ebeling-Witte, Frank, and Lester, 2007; Hamburger and Ben-Artzi, 2000; Scealy, Phillips, and Stevenson, 2002; Stritzke, Nguyen, and Durkin, 2004; Ward and Tracey, 2004), and the second to use the Internet and computer technologies as investigative laboratory tools (Sheeks and Birchmeier, 2007) to manipulate contextual cues in order to expand traditional FTF research (Brunet and Schmidt, 2007; Brunet and Schmidt, 2008). Both lines of research can provide useful implications for the understanding, intervention, and treatment of shyness and social anxiety. To date, the former line of research has been well-studied, although the latter is growing in popularity.

SHYNESS AND THE INTERNET: SURVEY STUDIES

The Internet is hypothesized to be more appealing to shy and socially anxious individuals as a medium of social communication due to the anonymity (i.e., lack of face-to-face contact) associated with computer-mediated communication (CMC). Unlike FTF, body language, voice tone, attractiveness, as well as other FTF issues are irrelevant online. Therefore, the anonymous online setting should lessen the heightened self-awareness and self-evaluative thoughts experienced by shy individuals during social interactions. It has been hypothesized that during online conversations, shy individuals will act more boldly than they typically do in FTF conversations (Roberts, Smith, and Pollock, 2000) and will use the Internet more frequently as a social medium than their non-shy peers in order to meet their relationship needs (McKenna and Bargh, 1999).

McKenna and colleagues (Bargh, McKenna, and Fitzsimons, 2002; McKenna, Green, and Gleason, 2002) demonstrated that socially anxious individuals are more comfortable presenting their 'real' me online than their peers who do not require CMC to do so as they adequately present their 'real' me in FTF situations. The concept of 'real' me is adapted from the true self presented in Carl Roger's theory of personality. Rogers (1951) postulated that true self is who you are when you are free of perceived constraints imposed on you by others. McKenna and colleagues also reported that it is a benefit for socially anxious individuals to be able to express their true self online (Bargh et al., 2002; McKenna et al., 2002). Further evidence supports that introverts are also better able to express their 'real' me online, whereas extraverts are better able to do so in traditional FTF settings (Amichai-Hamburger et al., 2002), and that extraversion in women is negatively related to using the Internet for social purposes (Hamburger and Ben-Artzi, 2000. These studies lend support to the idea that using the Internet as a social medium is more appealing to individuals who are shy so that they may express themselves, and less appealing to non-shy individuals, because they already express themselves well in FTF settings.

Others report that despite the frequency of shy individuals' Internet use for social communication, non-shy users are more at ease than shy users during online social interactions. Ward and Tracey (2004) conducted a self-report study with 414 participants. They asked the participants if they were currently or previously been involved in an online relationship (platonic or romantic) with someone they have never met in person. The participants who experienced an online relationship scored higher on shyness than did the individuals who had not. Shy individuals reported greater relationship involvement in their online friendships compared to their online friendship (e.g., more satisfaction with their online friendships than their offline friendships). However, shyness was still negatively correlated with relationship involvement measures, suggesting that non-shy individuals are more at ease with online relationships than shy individuals. Similarly, in a self-report study, a higher percentage of the high shy participants (73%) reported joining an Internet dating website to find a romantic or sexual relationship compared to the low shy participants (46%) (Scharlott and Christ, 1995). Similarly to the 'real' me studies (Amichai-Hamburger et al., 2002; Bargh et al., 2002; McKenna et al., 2002), shy participants were more likely to report being able to explore their personality and fantasies online than their non-shy counterparts (Scharlott and Christ, 1995). However, the high shy individuals sent fewer messages to other people through the Internet dating website than the low shy group. The studies by Scharlott and Christ (1995) and Ward and Tracy (2004) both demonstrate that shy individuals do feel more comfortable than non–shy adults communicating via the Internet than FTF interactions yet still exhibit some timid behaviour.

Despite using the Internet more than non-shy individuals, shy individuals use the Internet in a more hesitant manner than non-shy individuals. In a survey study of 134 university students, Stritze, Nguyen, and Durkin (2004) examined rejection sensitivity (i.e., the anxious expectation of rejection), initiation of relationships, self-disclosure, and emotional support in both offline and online settings. They found a context by shyness interaction on the first three measures, but not for emotional support. In offline settings, shy individuals and their non-shy peers reported significantly different levels of rejection sensitivity, initiation of relationships, and self-disclosure. In online settings, shy participants reported the same levels on these measures as the non-shy participants. Both groups reported feeling less shy in online settings, but the difference was more dramatic for the shy group. Their online reported shyness was still higher than that of the non-shy group. This study supports the idea that shy individuals are more comfortable online than offline, but does not support the idea that non-shy individuals would be more comfortable offline.

In contrast to the studies reporting that shy individuals use the Internet more for social communication, others failed to find such personality patterns in Internet users. Peris and colleagues (2002) compared a group of undergraduate students to a group of participants recruited from the Internet. They found no differences between the groups on extraversion and psychoticism. They concluded that the Internet users did not exhibit any specific personality pattern. Instead, they are typical individuals who merely use the Internet as another social medium. In line with these findings, Scealy, Philips, and Stevenson (2002) found that shyness did not predict Internet use for communicative purposes (e.g., email and chat rooms) in a sample of 300 participants. Although the findings of the reviewed articles could be interpreted as contradictory, there is, however, an underlining pattern that can be found in all of these studies. Shy individuals (or introverted and socially anxious individuals) appear to act more boldly and be more comfortable engaging in social interactions online than to traditional FTF. Nonetheless, non-shy individuals also use the Internet for the same purposes. It appears that the Internet "levels" the playing field so that shy individuals can be just as (or at least closer to being as) outgoing as non-shy individuals, allowing both groups to use the Internet as a social medium (Birnie and Horvath, 2007).

SHYNESS AND THE INTERNET: LABORATORY STUDIES

There is a long and rich history of FTF research demonstrating the importance of the role of context. Bem and Allen (1974) postulated that some of the time people and their behaviour can be predicted by personality traits, whereas other times they are predicted by situational variables (i.e., context). Despite the importance of their findings, they were based solely on self-reported behaviours, and reports from parents and friends. They did not objectively observe people in different situations. Additionally, context was traditionally defined as differences in situations (e.g., school versus home setting). Today, context has been examined on more micro-levels than in the past, and context has begun to be divided into its elements and cues (e.g., sex of conversation partner). A recent study demonstrated that males disclose less emotional information to other men, whereas sex of conversation partner does not influence women's emotional self-disclosure (Burleson, Holmstrom, and Gilstrap, 2005). Using the Internet as a research tool, it is possible to systematically control additional contextual cues that cannot be controlled as easily in FTF settings (e.g., visual cues, identity of conversation partner's sex, turn taking). This second line of research involving the Internet is, albeit less common, more time and labour intensive, but nonetheless, an informative objective method that can be used to control a maximum number of contextual cues.

As noted earlier, studying the nature and frequency of Internet use can provide insightful findings about the influence of individual differences in personality on CMC behaviour. However, these studies were mostly reliant on self-reported behaviour. For example, in Stritze and colleagues' (2004) study, participants self-reported their amount of self-disclosure for online and offline settings. Their self-disclosure was not objectively measured. Using the Internet as a tool, researchers could extend these findings by examining the micro-elements of the context and manipulate them to address, for example, two related types of issues. One issue comprised differences in social communicative behaviour expressed in CMC settings and FTF settings (i.e., typing versus talking). A second issue focuses on using the Internet's features (e.g., video and auditory broadcasting capabilities) to control more contextual cues than possible in FTF settings. The extra control afforded by the Internet as a tool is that it allows for more reliable results than tradition FTF studies.

Some researchers have begun to use the Internet as a research tool. Joinson (2001) conducted a series of studies utilizing the Internet to control visual cues to vary anonymity. In all three studies, he paired participants to form dyads. In Study 1, the dyads were randomly assigned to either a FTF condition or to a CMC condition and given a dilemma to discuss. In Study 2, Joinson (2001) controlled the level of visual anonymity by creating two CMC conditions. One condition (i.e., the visually anonymous condition) was exactly the same as the CMC condition from Study 1: participants discussed the dilemma via a chat program. Allowing the participant to see one another via a video conferencing system created the second CMC condition (i.e., the non-anonymous condition). In the final study, Joinson (2001) manipulated private versus public self-awareness by using the video-broadcasting feature of the chat program. To heighten private self-awareness, the video-feed was broadcasted on the participant's own computer screen but not on the partner's monitor. Conversely, in the heightened public self-awareness, the participants were instructed that their conversation partner and the experimenter would be able to see them via the video broadcasting but they could not see themselves on the monitor. The series of studies allowed Joinson (2001) to systematically control levels of visual cues to manipulate visual anonymity and self-awareness by using the Internet as a research tool, which would have been difficult in traditional FTF laboratory conditions. Joinson's protocol could easily be replicated to address the effects of personality, in particular shyness, in combination with the effects of varying levels of visual anonymity have on social communicative behaviour.

To investigate individual difference characteristics such as sex differences, Postmes and Spears (2002) created quartets (2 males and 2 females) and used Internet and computer technologies as their controlled laboratory environment. The groups were assigned a dilemma to discuss via synchronous CMC (i.e., similar to a chat room). They found no sex differences for number of turns, number of words, number of questions, but found that men made more autonomous statements than women. For this specific study, gender was identifiable by the chat room name that each participant had (e.g., woman1, man2).

In another study examining sex differences and Internet/computer technologies, Thomson (2006) chose to keep sex unidentifiable by using a similar Internet set up. Quarters of 2 men and 2 women were given 4 topics to discuss (2 male stereotypical topics and 2 female stereotypical topics). The chat room names were not sex specific to avoid identifying the sex of each participant. Female participants disclosed more personal information, asked more questions, and referred to other participants' messages more than did the male participants. During female-stereotypical topics, participants used more female-preferential language (e.g., self-derogatory comments, and agreeing more with others). Similarly, for male-stereotypical topics, male-preferential language was used more (e.g., opinions and insults). These two studies (Postmes and Spears, 2002; Thomson, 2006) demonstrate how similar studies may be designed, but with controlling one contextual cue depending on the goal of the experiment. By using the Internet, contextual cues (e.g. sex) can now be controlled by the experimenters. To date, there have been, however, few studies using the Internet as a tool to investigate how context mediates the influence of shyness on social communicative behaviour.

We have recently conducted two studies to examine how visual anonymity moderates the influence of shyness on social communicative behaviour (Brunet and Schmidt, 2007; Brunet and Schmidt, 2008). In these studies, we paired participants (female undergraduate students) to form dyads who never met in person throughout the entirety of the experiment. Each participant was in an isolated testing room equipped with an Internet accessible laptop. We established two conditions differing only in visual anonymity, which are replications of Joinson's (2001) two CMC groups: the visually anonymous condition and the visual non-anonymous condition. To create these conditions, we used webcams. The participants in the visual non-anonymous condition could see and be seen by their partner via the webcams, whereas the participants in the visually anonymous condition were chatting without the benefit of seeing one another via the webcams. The protocol of the experiment was set up to mimic a natural online conversation. Participants were instructed to get to know one another by chatting via an instant messenger for 10 minutes to chat in the first study (Brunet and Schmidt, 2007), and 20 minutes in the second study (Brunet and Schmidt, 2008). After the experiment, participants completed personality questionnaires, including the Cheek and Buss Shyness and Sociability scales (Cheek and Buss, 1981).

In addition to the benefit of being able to control visually anonymity without changing any other aspect of the experiment, using instant messengers as a tool allowed for perfect transcription of the conversation. Instead of the labour and resource intensive task of transcribing FTF conversations verbatim, all of the messages exchanged between the participants were recorded and saved automatically. The ultimate goal of the studies was to extend the FTF literature on the influence of shyness on social communicative behaviour. Therefore, our hypotheses were guided by the findings described in the *Shyness and Social Communicative Behaviours in FTF settings* section of this review paper.

In Study 1, we coded for 3 categories of measures: conversational style measures (e.g. self-disclosures), traditional linguistic measures (e.g., number of words per turn), and Internet-specific measures (e.g., emoticons) (Brunet and Schmidt, 2007). We found that shyness was negatively related to the number of self-disclosures made by participants in the visually non-anonymous condition, but no such relation was found in the visually anonymous condition.

In Study 2, we also examined a fourth set of measures namely affective language use measures (e.g., proportion of active words) (Brunet and Schmidt, 2008). Except for the Internet-specific measures, all of the measures can be coded in FTF transcripts as well. We selected extremely high shy participants and paired them with selected extremely low shy participants. In line with our findings from Study 1, the high shy participants significantly disclosed less personal information in the visually non-anonymous condition than the low shy participants, but no differences were found in the visually anonymous condition. Therefore, we were able to objectively measure self-disclosure while controlling visual anonymity. By doing so, we contributed to two areas of shyness research. The first is that we provide evidence supporting the hypothesis that shy individuals act more boldly online, but only when visual anonymity is maintained. We also contributed to the FTF research on shyness, by demonstrating that the effects of shyness on self-disclosure, as previously reported (Matsushima et al., 2000; Meleshko and Alden, 1993), are influenced by contextual cues such as visual anonymity. This result could imply that the self-consciousness and self-preoccupation that shy individuals experience during social interactions is not due to interaction itself, but instead to the exchange of visual cues.

Our findings also demonstrated that the effects of shyness on some forms of behaviour remain consistent despite differences in visual anonymity. For example, in our second study, shy individuals initiated fewer topics throughout the conversation than did the low shy participants regardless of condition. Likewise, their use of affective language remained the same regardless of condition (e.g., shy individuals used less fun words than their low shy partners).

Conclusions, Implications, and Future Directions

Our study design is only one of the many possible experimental designs for using the Internet as a research tool to study shyness, social anxiety, and social phobia (Brunet and Schmidt, 2007; Brunet and Schmidt, 2008). For example, researchers could modify our design to replicate Joinson's (2001) third study. Currently, we replicated the conditions Joinson (2001) established in his second study (i.e., visually anonymous versus visually non-anonymous). Researchers could manipulate the private versus public self-awareness levels of the conversations by having a condition in which the participants could see themselves via the webcam. With our current design, the participants could only see their partner. Considering that shy individuals have heightened self-consciousness and self-preoccupation (Crozier,

1979; Fatis, 1983), it seems plausible to predict that a condition high in both private and public self-awareness would result in high behavioural inhibition of shy individuals, and that a condition low in both (i.e., the visual anonymous condition) would allow shy individuals to be bolder.

With additional control of contextual cues by using the Internet as a laboratory tool, we can more reliably pinpoint exactly how shyness and other personality traits influence behaviour, especially social communicative behaviour than in FTF situations. With a firmer grasp of how shy individuals' heightened self-perception and avoidance behaviours are elicited, research can be directed to developing effective intervention and treatment options to help overcome the effects of shyness, social anxiety, and social phobia. The Internet could then be used not only as an experimental tool, but also as an intervention tool.

ACKNOWLEDGMENTS

This work was supported by a grant from the Social Science and Humanities Research Council (SSHRC) of Canada awarded to Louis A. Schmidt.

REFERENCES

Amichai-Hamburger, Y., Wainapel, G., and Fox, S. (2002). "On the Internet no one knows I'm an introvert": Extroversion, neuroticism, and Internet interaction. *CyberPsychology and Behavior, 5*, 125-128.

Bargh, J. A., McKeena, K. Y. A., and Fitzsimons, G. M. (2002). Can you see the real me? Activation and expression of the "true self" on the Internet. *Journal of Social Issues, 58*, 33-48.

Bem, D. J., and Allen, A. (1974). On predicting some of the people some of the time: The search for cross-situational consistencies in behavior. *Psychological Review, 81*, 506-520.

Birnie, S. A. and Horvath, P. (2002). Psychological predictors of Internet social communication. *Journal of Computer-Mediated Communication, 7*. Retrieved from http://jcmc.indiana.edu/vol7/issue4/horvath.html

Brunet, P. M., and Schmidt, L. A. (2007). Is shyness context specific? Relation between shyness and online self-disclosure with and without a live webcam in young adults. *Journal of Research in Personality, 41*, 938-945.

Brunet, P. M., and Schmidt, L. A. (2008). Context influences computer-mediated self-disclosure but not the use of affective language in shy adults. *Manuscript submitted for publication*.

Burleson, B. R., Holmstrom, A. J., and Gilstrap, C. M. (2005). "Guys can't say that to guys": Four experiments assessing the normative motivation account for deficiencies in the emotional support provided by men. *Communication Monographs, 72*, 468-501.

Chak, K., and Leung, L. (2004). Shyness and locus of control as predictors of Internet addiction and Internet use. *CyberPsychology and Behavior, 7*, 559-570.

Cheek, J. M., and Buss, A. H. (1981). Shyness and sociability. *Journal of Personality and Social Psychology, 41*, 330-339.

Crozier, R. (1979). Shyness as anxious self-preoccupation. *Psychological Reports*, *44*, 959-962.

Ebeling-Witte, S., Frank, M. L., and Lester, D. (2007). Shyness, Internet use, and personality. *Cyberpsychology and Behavior*, *10*, 713-716.

Fatis, M. (1983). Degree of shyness and self-reported physiological, behavioral, and cognitive reactions. *Psychological Reports*, *52*, 315-354.

Garcia, S., Stinson, L., Ickes, W., Bissonnette, V., and Briggs, S. R. (1991). Shyness and physical attractiveness in mixed-sex dyads. *Journal of Personality and Social Psychology, 61*, 35–49.

Hamburger, Y. A., and Ben-Artzi, E. (2000). The relationship between extraversion and neuroticism and the different uses of the Internet. *Computers in Human Behavior, 16*, 441-449.

Hill, G. J. (1989). An unwillingness to act: Behavioral appropriateness, situational constraint, and self-efficacy in shyness. *Journal of Personality, 57*, 871-890.

Joinson, A.N. (2001). Self-disclosure in computer-mediated communication: The role of self-awareness and visual anonymity. *European Journal of Social Psychology, 31*, 177-192.

Kagan, J. (1994). *Galen's Prophecy*. New York: Basic Books

Manning, P., and Ray. G. (1993). Shyness, self-confidence, and social interaction. *Social Psychology Quarterly, 5*, 178-192.

Matsushima, R., Shiomi, K., and Kuhlman, D. M. (2000). Shyness in self-disclosure mediated by social skill. *Psychological Reports, 86*, 333-338.

McKenna, K. Y. A., and Bargh, J. A. (1999). Causes and consequences of social interaction on the Internet: A conceptual framework. *Media Psychology, 1*, 249-269.

McKenna, K. Y. A., Green, A. S., and Gleason, M. E. J. (2002). Relationship formation on the Internet: What's the big attraction? *Journal of Social Issues, 58*, 9-31.

Meleshko, K.G.A., and Alden, L.E. (1993). Anxiety and self-disclosure: Toward a motivational model. *Journal of Personality and Social Psychology*, *64*, 1000-1009.

Pennebaker, J. W. and King, L. A. (1999). Linguistic styles: Language use as an individual difference. *Journal of Personality and Social Psychology*, *77*, 1296-1312.

Peris, R., Gimeno, M. A., Pinazo, D., Ortet, G., Carrero, V., Sanchiz, M. et al. (2002). Online chat rooms: Virtual spaces of interaction for socially oriented people. *CyberPsychology and Behavior, 5*, 43-51.

Pilkonis, P. A. (1977). The behavioral consequences of shyness. *Journal of Personality,45*, 596-611.

Postmes, T., and Spears, R. (2002). Behavior online: Does anonymous computer communication reduce gender inequalities? *Personality and Social Psychology Bulletin, 28*, 1073-1083.

Roberts, L. D., Smith, L. M., and Pollock, C. M. (2000). U r a lot bolder on the net. In W. R. Crozier (Ed.), *Shyness, development, consolidation, and change* (pp. 121-138). Boston: McGraw-Hill.

Rogers, C. (1951). *Client-centered therapy*. Boston: Houghton-Mifflin.

Scealy, M., Phillips, J. G., and Stevenson, R. (2002). Shyness and anxiety as predictors of patterns on Internet usage. *CyberPsychology and Behavior, 5*, 507-515.

Scharlott, B. W., and Christ, W. G. (1995). Overcoming relationship-initiation barriers: The impact of a computer-dating system on sex role, shyness, and appearance inhibitions. *Computers in Human Behavior, 11*, 191-204.

Sheeks, M. S., and Birchmeier, Z. P. (2007). Shyness, sociability, and the use of computer-mediated communication in relationship development. *CyberPsychology and Behavior, 10*, 64-70.

Stritzke, W. G. K., Nguyen, A., and Durkin, K. (2004). Shyness and computer-mediated communication: A self-presentational theory perspective. *Media Psychology, 6*, 1-22.

Taylor, J. (2002). Introduction. [Special Issue: Psychology and the Internet]. *Social Science and Computer Review*, 233-240.

Thomson, R. (2006). The effects of topic of discussion on gendered language in computer-mediated communication discussion. *Journal of Language and Social Psychology, 25*, 167-178.

Thompson, S., and Rapee, R. M. (2002). The effects of situational structure on the social performance of socially anxious and non-anxious participants. *Journal of Behavior Therapy and Experimental Psychiatry, 33*, 91-102.

Ward, C. C., and Tracey, T. J. G. (2004). Relation of shyness with aspects of online relationship involvement. *Journal of Social and Personal Relationships, 21*, 611-623.

Zimbardo, P. G. (1977). *Shyness: What is it, what to do about it*. Reading, MA: Addison-Wesley.

In: Computer Networks, Policies and Applications
Editors: C. M. Berger and A. D. Streusse
ISBN: 978-1-61209-088-7

Chapter 11

ONLINE BUT OFF-BALANCE: GENDER DIFFERENCES IN HIGH SCHOOL INTERNET COVERAGE

Bryan Butler and Michael Sagas
Texas A and M University, USA

ABSTRACT

More women and girls are participating in athletics than ever before, however, media coverage of female athletes still lags behind that of male athletes (Pederson and Whisenant, 2002). The purpose of this study was to continue to assess media coverage by gender at the high school level by assessing the coverage of school websites. An initial check at the midpoint of the season evaluated six items, including the presence of coaches' names and rosters. Data from eight dates were then collected for the rest of the season to assess game results posted online. Results indicated that males received more coverage than females in general information and in game results. The findings point to potential Title IX violations by high schools.

Since the introduction of Title IX, more girls are participating in sports than ever before (Braddock, Sokol-Katz, Greene, and Basinger-Fleischman, 2005). Studies have shown that there are benefits for girls who participate in sports that go beyond improved physical fitness. In the classroom, girls who participate in sports are less likely to have unexcused absences from school. They are more likely to take core courses and have higher grade point averages. And they have higher expectations on attending college after high school (Videon, 2002). Outside of class, high school girls who participate in high-performance, competitive sports appear less likely to engage in sexual risk-taking (Savage and Holcomb, 1999). Yet, despite the benefits, girls' participation in sports still lags behind the participation rate for boys (Braddock et al., 2005).

Shakib and Dunbar (2002) wrote that attempts to eliminate what they called the "patriarchy within sport" need to focus on two parts: (a) structural factors, using laws such at Title IX, and (b) ideological factors, looking for ways to break down limiting gender expectations. Providing equal opportunities and coverage for female athletes allows public schools at all levels to comply with the law while at the same time working to give female athletes a sense of equality in sport.

Media coverage of female athletes also trails the coverage of male athletes (Lumpkin and Williams, 1991). Many of the studies of gender and media show that even female professional and college athletes receive less coverage than males who compete at similar levels (Lumpkin and Williamson, 1991; Sagas, Cunningham, Wigley, and Ashley 2000; Shifflett and Revelle, 1994; Wann, Schrader, Allison, and McGeorge, 1998). Similar findings appear to hold true at the high school level. Pederson and Whisenant (2002) found in their study of Florida newspapers that coverage of high school sports was more likely to be about males than females. But smaller newspapers were more likely to provide coverage of female high school athletics (40 percent of their high school coverage) than medium (33 percent) or large (32 percent) newspapers.

Media also can have an effect on girls and how they see themselves (Polce-Lynch, Myers, Kliewer, and Kilmartin, 2001). Yet, when covering sports or producing features on teams or athletes, media many times choose to focus on the physical prowess or accomplishments of males while showing females in supporting roles or focusing on their physical appearances rather than showing what female athletes have achieved (Lumpkin and Williams, 1991; Cuneen and Sidwell, 1998; Knight, 2001; Messner, Duncan and Cooky, 2003).

Most media are private entities that require revenue to remain in business. They are essentially free to cover sports in the manner in which they feel is best for their businesses. Public high schools, on the other hand, are bound by Title IX, the same as public colleges and universities (Braddock et al., 2005). Public high schools are required to provide equal opportunities to girls in the classroom and on the field. Equivalent treatment should be standard in all areas, including the amount of information provided on high school athletics websites. Using Title IX as a way to ensure equal opportunities at the high school level has become more common. According to Pennington (2004) on Title IX, "the number of complaints involving sex discrimination in high school and even middle school athletics has outpaced those involving colleges by five to one since 2001."

At the college level, however, studies have shown that equality in coverage has yet to be achieved. Sagas et al. (2000) studied the school athletics Web sites of 52 NCAA Division I universities, comparing baseball and softball coverage. The researchers found more information was made available about baseball and male athletes than for softball and female athletes, both in initial information, such as coaches' names or schedule listings, and in results made available during the season.

This exploratory study looks at high school athletic websites. Because high schools vary in the amount of resources they have to work with, it is likely that there is great variation between schools in the amount of information they provide. However, the amount of coverage for male and female athletes should be the equivalent each school. This study concentrated on basketball, which has the same number of players on the court at any one time, whether males or females are playing, and thus equal coverage is expected.

Methods

Six different Texas high schools were selected from a list of schools known to have athletics Web sites. The study looked at boys and girls basketball during the season and into postseason play. An initial check of the sites took place Jan. 26, after the season had started, and looked for six different preliminary items for the boys and girls basketball teams, including coaches' names, rosters and players' pictures. Each item was rated as either present or absent on the site.

During the season, the sites were randomly checked twice each week, once each on Mondays and Wednesdays, the first school days after regularly scheduled games. Five items were checked, four for present or absent, and one counted the number of words for each press release. This part of the study also focused on whether results for males and females were posted in a timely manner on the websites, so updates to previous games that occurred between checks were not included in the data. Two coders independently coded the first week and preliminary data. The results of the coding indicated 100% agreement between the two coders, thus a single coder collected data for the remainder of the study.

Results

In most areas, male athletes received more coverage than females. Frequency analyses indicated that males received more coverage in three categories at the midpoint check, similar coverage in two categories, and girls received more coverage in one (see Table 1).

Table 1. Variables Analyzed for Boys' and Girls' Basketball Sites During the Season

Measure	Boys %	Girls %	χ^2
Results posted	61.0	46.2	4.044*
Statistics posted	28.6	7.7	11.685*
Pictures posted	4.8	0.0	3.308*
Press release posted	28.6	12.8	6.051*

Note. p's < . 01

Table 2. Variables Analyzed for Boys' and Girls' Basketball Sites for Static Information

Measure	Boys %	Girls %	X^2
Head coach's name posted	100	100	___*
Head coach's picture posted	17.6	0.0	.296
Assistant coach's name posted	83.3	100	.296
Assistant coach's picture posted	83.3	100	.296
Roster posted	100	100	___*
Player pictures posted	33.3	16.7	.505*

Note. p's < . 01.

Results from chi square analyses of the weekly website assessments indicated that the boys received between 4.8% - 20.9% greater coverage across each category during the season (all *p*'s < . 05, see Table 2). Results from the t-test comparing the length of coverage offered in weekly releases also indicated that the boys basketball teams received longer coverage in the number of words (M = 58.62, SD = 158.12) than the girls teams (M = 14.87, SD = 56.99); $t\,(160) = 2.38, p < .05$. The findings are similar to recent literature related to media coverage in high school athletics.

Discussion

More than 30 years after the adoption of Title IX, the issues of fairness and equal opportunity for female athletes have not been resolved. Despite the benefits athletic participation seem to provide girls (Videon, 2002; Savage and Holcomb, 1999), many times they are not given equal access or treatment. This study into high school athletic websites was to investigate whether inequity might be present.

The "patriarchy within sport" offered by Shakib and Dunbar (2002) appears to still have a hold. Despite Title IX and the legal ramifications it holds for schools, this study found that equivalent information wasn't provided for female student-athletes at the high schools in this study. It might be difficult to change the ideological factors that hold sway when looking for ways to break down limiting gender expectations when laws appear to be ineffective in changing the ways schools function.

Also, if media does indeed effect how girls see themselves (Polce-Lynch, Myers, Kliewer, and Kilmartin, 2001), then perhaps equal coverage of females being success in athletics would provide more positive role models. Schools that provide statistics, records and scores help change the patriarchy by not showing female athletes in strictly a supporting role. Providing the same type of coverage for male and female athletes would lessen the objectification of female athletes that other researchers have found in traditional media (Lumpkin and Williams, 1991; Cuneen and Sidwell, 1998; Knight, 2001; Messner, Duncan and Cooky, 2003).

High school websites are maintained by different sources. It appears from this study that some high schools do the work themselves, while others allow booster clubs or third parties maintain athletics websites. In the cases in which the high schools maintained their own sites or had a third party maintain the site, the websites were set up for equal coverage for males and females. When booster clubs were involved, the sports that booster clubs maintain sites many times had more elaborate coverage than sports that did not have booster clubs maintaining their sites. Further research needs to look into the role of booster clubs and whether they affect coverage in male and female sports, and whether this also represents potential Title IX violations.

Looking at websites, it is impossible to determine where a breakdown in coverage might occur. Results and information must pass from the coach or scorekeeper on to the web designer or maintainer to be posted. Then the person in charge of the site must make the needed changes. In this process, there are different ways for scores and results to fall through the cracks and not make it online. However, at any particular school, the process involved would seem to be the same to post results from all sporting events. Therefore, it would seem that results from male or female sporting events should be equally plausible to not be posted.

This study was exploratory in nature, and the number of schools studied was small. An expanded study would provide a clearer picture of trends in athletic websites and the coverage they provide. This study also focused on whether results were posted in a timely manner on the websites, usually the day after contests were scheduled. It is possible that some sites did provide more coverage for female athletics later than a day or two after the event. However, results for males were collected in identical fashion, and so the disparity in coverage should be the same for both.

Public high schools are bound by Title IX to provide equal treatment for male and female athletes. Until schools take steps to ensure female athletes receive equal treatment, they will continue to find themselves in the ethical low ground and open to legal action.

References

Braddock II, J.H., Sokol-Katz, J., Greene A., and Basinger-Fleischman, L. (2005). Uneven playing fields: State variations in boys' and girls' access to and participation in high school interscholastic sports. *Sociological Spectrum*, 25 (2), 231-250.

Cuneen, J., and Sidwell, M.J., (1998). Gender portrayals in *Sports Illustrated for Kids* advertisements: A content analysis of prominent and supporting models. *Journal of Sport Management*, 12, 39-50.

Knight, J.L., and Giuliano, T.A., (2001). He's a Laker; she's a looker: The consequences of gender-stereotypical portrayals of male and female athletes by the print media. *Sex Roles: A Journal of Research*, 45 (3-4), 217-229.

Lumpkin, A., and Williams, L.D., (1991). An analysis of *Sports Illustrated* feature articles, 1954-1987. *Sociology of Sport Journal*, 8, 16-32.

Messner, M.A., Duncan, M.C., and Cooky, C., (2003). Silence, sports bras, and wrestling porn: Women in televised sports news and highlights shows. *Journal of Sport and Social Issues*, 27 (1), 38-51.

Pedersen, P.M. and Whisenant, W.A., (2002). Amount of newspaper coverage of high school athletics for boys and girls on sports page and newspaper circulation. *Perceptual and Motor Skills*, 94, 323-326.

Pennington, B., (2004, June 29). Title IX trickles down to girls of Generation Z. *The New York Times*, pp. D1, D2.

Polce-Lynch, M., Myers, B.J., Kliewer, W., and Kilmartin, C., (2001). Adolescent self-esteem and gender: Exploring relations to sexual harassment, body image, media influence, and emotional expression. *Journal of Youth and Adolescence*, 30 (2), 225-244.

Sagas, M., Cunningham, G.B., Wigley, B.J., and Ashley, F.B., (2000). Internet coverage of university softball and baseball Web sites: The inequity continues. *Sociology of Sport Journal*, 17, 198-205.

Savage, M.P., and Holcomb, D.R. (1999). Adolescent female athletes' sexual risk-taking behaviors. *Journal of Youth and Adolescence*, 28 (5), 595-602.

Shakib, S., and Dunbar, M.D., (2002). The social construction of female and male high school basketball participation: Reproducing the gender order through a two-tiered sporting institution. *Sociological Perspectives*, 45 (4), 353-378.

Shifflett, B., and Revelle, R., (1994). Gender equity in sport media coverage: A review of the *NCAA News*. *Journal Of Sport and Social Issues*, 18 (2), 44-150.

Videon, T.M. (2002) Who plays and who benefits: Gender, interscholastic athletics, and academic outcomes. *Sociological Perspectives*, 45 (4), 415-444.

Wann, D.L., Schrader, M.P., Allison, J.A., and McGeorge, K.K., (1998). The inequitable newspaper coverage of men's and women's athletics at small, medium, and large universities. *Journal of Sport and Social Issues*, 22 (1), 79-87.

INDEX

9

A

B

C

D

G

H

I

J

K

L

M

N

O

P

Q

T

U

V